Out of This World
NEW MEXICO'S CONTRIBUTIONS TO SPACE TRAVEL

Loretta Hall

published by Rio Grande Books

All rights reserved. Rio Grande Books, an imprint of LPD Press
        Los Ranchos, New Mexico
        www.LPDPress.com

Printed in the U.S.A.

Book design by Paul Rhetts
Cover illustration: Courtesy of Virgin Galactic

Library of Congress Cataloging-in-Publication Data

Hall, Loretta.
Out of this world : New Mexico's contributions to space travel / by Loretta
        Hall.
        p. cm.
        ISBN 978-1-890689-79-7 (pbk. : alk. paper)
        ISBN 978-1-890689-86-5 (hardcover : alk. paper)
    1. Astronautics--Research--New Mexico--History. I. Title. II. Title: New
        Mexico's contributions to space travel.

TL789.8.U5.H28 2011
    629.4072--dc22

2011008960

Table of Contents

Prologue: 1598—The Road to the Future Begins 5

Chapter One: Goddard's Rockets 9

Chapter Two: Operation Paperclip & Beyond 27

Chapter Three: Animals Aloft 45

Chapter Four: Silent Skyhooks 63

Chapter Five: The Human Factor 85

Chapter Six: Window on the Universe 105

Chapter Seven: Alien Visitors 121

Chapter Eight: Xpediting Commercialization...........................135

Chapter Nine: Spaceport America 151

Epilogue: Critical Mass 169

Photo Credits...171

Index...173

Acknowledgments

As with many history projects, much of my research for *Out of this World* involved poring through numerous documents, some well known and some relatively obscure. I am grateful to the people who wrote them and left records of the astonishing adventures I described in this book. I am also grateful to the following people who shared their time, knowledge, and resources with me in person: Buzz Aldrin, former NASA astronaut and founder of the ShareSpace Foundation; David A. Clary, author of *Rocket Man: The Life and Legends of Robert H. Goddard, American Pioneer of Space Flight*; Rick Homans of the New Mexico Spaceport Authority; George House, Wayne Mattson, and Michael Shinabery at the New Mexico Museum of Space History; Joe Kittinger, the first person to parachute from space; and Laurie Rufe and Candace Russell at the Roswell Museum and Art Center.

Doing the research and writing the manuscript are only the first steps in producing a book. I appreciate the efforts of these people who helped get this one into print: Bridget Grumet and Bernadette Bruha, who reviewed the manuscript and offered suggestions for improving it, and Paul Rhetts and Barbe Awalt of Rio Grande Books, who fashioned the raw manuscript into a real book.

Finally, I want to thank my husband, Jerry Hall, for his enthusiastic support and enouragment.

Foreword

In early August 2010, I was in line to board at the El Paso, Texas, airport when I realized that standing quietly next to me was Joe Kittinger. Just days away would be the fiftieth anniversary of his record-breaking jump above southern New Mexico, when he proved that astronauts could survive a high-altitude ejection. In his career, Kittinger, along with others around New Mexico over the past sixty-five years, performed extraordinary feats in the first efforts to reach space. Many of those pioneers are gone, but some, such as Kittinger, live among a public that has no idea of their accomplishments.

Until, that is, today. Loretta Hall's *Out of this World* has brought together their tales in a timeline that leaves no doubt why New Mexico has been, and continues to be, a leader in the business of space. Thinking of Ms. Hall's writing as the type to be catalogued and shelved away in an archive, awaiting a future historian to take it down and blow off the dust, would be a grave mistake. *Out of this World* is not just for aficionados of space, but also for those interested in any history. Her telling reads like a thriller, with humor that will amuse any reader. Her sidebars contain little-known, but revealing and fascinating bits of trivia. She does not wallow in technical jargon, nor does she portray scientific pioneers such as Dr. Robert Goddard, the German Paperclip scientists, and Dr. John Stapp as "sticks-in-the-mud."

In the twenty-first century, one cannot get through a single day without benefiting from space. Do you take medicine, fasten a seat belt, or press *send* on a cell phone? Much of the day-to-day tools you take for granted evolved from space technology that was birthed in New Mexico. Ms. Hall fully illustrates how ancient residents in the days of prehistory viewed the heavens before the state's borders were drawn; and she does not shy away from the continuous UFO controversy, presenting both the "believer" and the "skeptic" viewpoints to allow readers to make up their own minds.

The detailed history of New Mexico's leadership in commercial space development gives almost a week-by-week development of who did what and why. Taken in context with the history, Ms. Hall shows why it is necessary to learn to walk before you can run, and why the state is rapidly approaching the running stage in space commercialization.

Loretta Hall's book is an important contribution to the history of space in New Mexico and should be read by all space enthusiasts, as well as future astronauts.

Michael Shinabery
New Mexico Museum of Space History Educational Specialist,
New Mexico Humanities Scholar
with
Wayne Mattson, USAF Lt. Col. (ret)
Archivist, New Mexico Museum of Space History

Prologue
1598—The Road to the Future Begins

In 1598, three European astronomers—Galileo Galilei, Johannes Kepler, and Tycho Brache—were helping each other develop the new astronomical theory that the planets revolve around the sun. They invented and refined instruments to track the planets' movements, developed mathematical models to predict eclipses, and compared the timing and duration of the actual events to their predictions. Their work created a foundation on which, three and a half centuries later, space travel could become a reality.

At the same time, half a world away, another event took place that would eventually create a stage for many of the developmental dramas of space travel. This event also involved traveling into new frontiers, but of a more rudimentary sort than rocket ships catapulting into the sky. Spain's King Phillip had authorized a Spanish conquistador named Juan de Oñate to colonize New Mexico. To accomplish that mission, Oñate had to forge a trail northward from New Spain (now Mexico) into the foreign territory. He began at the end of El Camino Real, an important trade route that stretched a thousand miles northward from Mexico City. The extension Oñate pioneered added 700 miles to the route, and it would remain the primary road into New Mexico for more than two centuries, until the Santa Fe Trail connected the territory to the United States. In fact, El Camino Real de Tierra Adentro (The Royal Road to the Interior) continued to be a vital transportation facility until a railroad crossed New Mexico in the 1880s.

Oñate began his pioneering expedition in the spring of 1598. It was a large caravan: more than 500 soldiers and settlers, along with 7,000 head of livestock and 83 oxcarts carrying supplies. Ten Franciscan friars came too, to convert the pagan Indians. Many of the travelers simply walked—that was the state of the art of transportation in the late sixteenth century. While traveling, the caravan could stretch out for

3 miles or more. The first 150 miles of their journey through the Chihuahuan desert from Santa Bárbara, Mexico, to the Rio Grande (Great River) took four months.

With the new territory of their dreams in sight across the broad river, the parched and exhausted colonists and animals stopped to recuperate and savor the plentiful water supply. While they rested, Oñate sent out scouts to find a place where the river was shallow enough to cross. Word of an acceptable spot soon came, on an auspicious day: April 30, 1598, the Feast of the Ascension, when Catholics celebrate the ascent into heaven of the risen Savior. The Franciscans celebrated Mass. The colonists, having regained their strength, held foot races and other sporting events. They even performed a play written by one of the soldiers, dramatizing how eagerly the Indians would receive them and embrace Christianity—and how the soldiers would decisively quash any resistance to colonization.

Several days later, the caravan crossed the Rio Grande and entered the newly claimed territory. Progress through the Mesilla Valley was relatively smooth, marred only by the occasional wagon breakdown or straying animal. After about three weeks, however, Oñate learned that, against his orders, a scouting party had revealed the expedition's existence to the Indians. He decided to push forward immediately with a quick-moving advance party of only sixty people to reassure the Indians of their peaceful intentions. They immediately met a severe geographical challenge: a 90-mile stretch of daunting desert that would become known as Jornada del Muerto (Journey of the Dead). For five arduous days, the party trudged across the desolate valley, finding no water, no shade, no firewood, and no vegetation for their horses to graze on. Finally, they happened upon a most welcome sight: the meandering Rio Grande. Even better, there was a Pueblo Indian village where they were fed and comforted. Oñate gave the village a new name: Socorro, the Spanish word for help.

The trail Oñate and his fellow pioneers forged from Mexico through the Jornada del Muerto and on to the new village of Santa Fe was one of the most important transportation achievements of North America. Four centuries later and 3 miles to the east, across the still-desolate landscape of Jornada del Muerto, another transportation revolution is taking place. The world's first purpose-built commercial space travel launching site, Spaceport America, is an unobtrusive but spectacularly innovative addition to the southern New Mexico desert. The access road to the spaceport actually crosses El Camino Real.

Spaceport America tenants, such as UP Aerospace, launch commercial and scientific payloads in unmanned rockets. Anchor tenant Virgin Galactic initially offers suborbital flights to space tourists. It plans to serve thousands of passengers a year with three flights a day, and it may eventually offer orbital flights as well.

Oñate's expedition charted new territory, but it did not revolutionize the means of transportation. Spaceport America does both. In accomplishing an astonishing extension to humanity's accessibility to outer space, Spaceport America seems to have burst forth from the mind of Zeus. In reality, however, it is simply the next chapter in an eight-decade-long series of programs, experiments, and incredible events tucked away in the unpretentious expanses of the New Mexico desert. From the first launch of animals (fruit flies) into space in 1947 until the first purpose-built commercial spaceport (offering even recreational space travel) in 2011, New Mexico has provided fertile soil for the growth of space travel for fun and profit.

Chapter One: Goddard's Rockets

Effie Ward's neighbors were frightened and furious. Bob Goddard had set off another one of his confounded rockets from the cabbage patch on his Aunt Effie's farm. This time, it was more than annoying, it was downright dangerous. They were not about to let him continue disrupting their peace and risking damage to their property. Massachusetts was not a proper place for these shenanigans.

He'd shot off the first one back in March of 1926. The rudimentary device—two liquid fuel tanks and a combustion chamber connected by a 10-foot-long, spindly skeleton of metal fuel lines—climbed 40 feet into the air and came down 2½ seconds later, 184 feet away from where it took off. A second rocket, launched a month later, stayed in the air for more than 4 seconds. Those were OK. After all, Bob—Dr. Robert Goddard—was a physics professor at Clark University in nearby Worcester, Massachusetts, so he must know what he was doing.

Goddard spent the next couple of years redesigning his rockets and testing them, without a successful liftoff. Sometimes the failures literally exploded in the launch tower, a modified windmill frame. Undaunted by the failures, which Goddard said provided valuable negative information, he kept tinkering. Then, the day after Christmas 1928, he had a successful firing, with the rocket reaching a speed of 60 miles an hour before hitting the ground 205 feet from the tower.

The first flight of a liquid-fuel rocket was as monumental as the first airplane flight had been, twenty-three years earlier. Goddard's rocket took 9½ seconds less to fly 11 feet higher and 85 feet farther. But the Wright Brothers' plane carried a person, and it flew a distance of 852 feet in 59 seconds later that same day. Maximum speeds: 60 miles an hour for the rocket and 34 miles an hour for the plane.

Aunt Effie's farm became a golf course in 1931. The Pakachoag Golf Course is now owned by the town of Auburn, Massachusetts. Midway between the tee and the green on the ninth fairway, a small granite obelisk marks the site of the world's first liquid-fuel rocket launch. It rests at the center of a 200-foot-radius circle designated as a National Historic Landmark.

But it was the next successful launch, on the afternoon of July 17, 1929, that infuriated the neighbors. The 11-foot-tall contraption rumbled in the tower for 13 seconds before starting to move upward. The exhaust flame grew to 20 feet as the rocket roared skyward. The rocket climbed for another 4 seconds, reaching a height of 90 feet before plummeting to the ground. When it crashed 170 feet from the tower, one of the fuel tanks exploded, setting the grass on fire.

That was more than enough to get the neighbors' attention. An article in the November 1929 issue of *Modern Mechanix* magazine put it this way: "His latest experiments caused a sensation around Worcester, where a group of villagers in the neighborhood of Goddard's experiment station observed what appeared to be a flaming meteor which hurtled through space at breathtaking gait, lighting up the landscape and finally bursting with a thunderous roar."

People as far as 2 miles away noticed the frightful event and called the police to report an airplane crash. As Goddard and his assistants finished retrieving their equipment, two ambulances and several cars rushed onto Aunt Effie's field to rescue victims of the supposed crash. In the days that followed, newspapers across the country carried the story with headlines like "Man in the Moon Scared Green" and "Moon Rocket Misses Target by 238,799½ Miles."

Citizen pressure and safety concerns led the State Fire Marshal to forbid Goddard from conducting any more rocket tests in Massachusetts.

Robert Goddard, the Man

Massachusetts was the only home Goddard had ever known. He was born in Worcester in 1882, but within a few months his family moved east to a suburb of Boston. When he was fifteen, his mother came down with tuberculosis and the family returned to Worcester in the center of the state. He went to college and graduate school in his hometown, earning a BS from Worcester Polytechnic Institute, and MA and PhD degrees in physics at Clark University. Other than a year as a research instructor at Princeton University, he worked in Worcester, teaching and doing research at Clark.

As a teenager, Goddard read a new novel that not only captured his imagination but held it hostage the rest of his life. The book, *The War of the Worlds* by H.G. Wells, vividly described an invasion of Earth by Martians. Many years later, following a successful launch that proved a new technology, he sent a letter to Wells. In it he wrote, "In 1898 I read your *War of the Worlds*. I was sixteen years old, and the new viewpoints of scientific applications, as well as the compelling realism of the thing, made a deep impression. The spell was complete about a year afterward, and I decided that what might conservatively be called 'high-altitude research' was the most fascinating problem in existence."

The event that made the spell complete started as an ordinary task. He climbed up into a cherry tree outside his family's house. Pausing from cutting dead branches, he looked out across the fields, and a magical idea came to his mind. "I imagined how wonderful it would be to make some device which had even the *possibility* of ascending to Mars, and how it would look on a small scale, if sent up from the meadow at my feet," he wrote in a 1927 autobiographical essay. That daydream became the compulsion that defined the rest of his life.

Eight years after his vision in the cherry tree, Goddard wrote an essay, "On the Possibility of Navigating Interplanetary Space," for his English class at Worcester Polytechnic Institute. In it, he wrote, "The discussion falls naturally into three divisions: the sustaining of life in space, the protection against accident during transit, and the means of propulsion." Addressing the first point, he noted that the problems of food storage, renewal of air in a closed space, and retention of warmth had already been solved. He found the protection issue troublesome, primarily because the British astronomer Sir Norman Lockyer predicted that meteors densely populated outer space—averaging separations of only 250 miles. Yet Goddard dismissed this problem with a paragraph, suggesting a fanciful notion of shielding the spaceship by having it travel within a meteor swarm headed in the desired direction. The swarm would deflect conflicting meteors, and any incidental contact with meteors within the storm would be minor because of the small relative velocities.

What really captured Goddard's attention was the third issue—the means of propulsion. In a dozen paragraphs, he expounded on the potentials of various energy sources—solar, chemical, and radioactive disintegration. By 1919, he was able to publish theoretical and experimental evidence that "with a rocket of high efficiency, consisting chiefly of propellant material, it should be possible to send small masses even to such great distances as to escape the earth's attraction."

Though his quest began with the notion of space travel, he learned not to talk about it in those terms. For one thing, it drew ridicule like the nickname "Moon Man." But from a more practical standpoint, he knew that developing a vehicle capable of reaching even the upper limits of the atmosphere was a monumental task. "How many more years I shall be able to work on the problem, I do not know; I hope, as long as I live," he wrote in his 1932 letter to Wells. "There can be no thought of finishing, for 'aiming at the stars,' both literally and figuratively, is a problem to occupy generations, so that no matter how much progress one makes, there is always the thrill of just beginning."

One of Goddard's students at Clark University was Edwin Aldrin, Sr., the father of astronaut "Buzz" Aldrin, the second man to walk on the moon.

A Suitable Site

With his rocket tests banned from the entire Commonwealth of Massachusetts, Goddard had to either give up or relocate. Developing a vehicle that could reach space was a mission—perhaps compulsion—that Goddard could not abandon. He would have to leave his home state and move to a more suitable place to continue his work. The question was, where?

The cloud of extravagant publicity about the July 1929 launch turned out to have a silver lining. Among the people who read newspaper and magazine accounts of the event were Charles Lindbergh and his friend, Harry Guggenheim. Lindbergh, whose unprecedented trans-Atlantic airplane flight in 1927 had made him a national icon, thought Goddard's inventions could revolutionize flight. Guggenheim, who had been a US Navy pilot during World War I, was president of the Daniel Guggenheim Fund for the Promotion of Aeronautics. Grants from individuals and foundations associated with the Guggenheim fortune would be a major source of funding for the rest of Goddard's career.

With a personal grant from Harry's father, Daniel Guggenheim, along with smaller ones from the Smithsonian and Carnegie Institutions, Goddard was financially able to move from Massachusetts and devote his full attention to his research. He prepared for the move by applying his usual scientific analysis to the selection of an appropriate location. First, he asked Lindbergh's opinion as a pilot with extensive experience flying over the entire country. Then he consulted a Clark University colleague, meteorology professor Charles Brooks.

"We sat down together and methodically combed over the weather statistics of various sections of the country," Goddard wrote in an unpublished article now held in the Clark University archives. "We wanted a relatively high region with a minimum of rain and snowfall, a minimum of cloudiness, and freedom from fog. We looked, too, for a place without extremes of heat and cold where we could count on considerable periods without wind. In other words, we wanted good outdoor working weather the year round, and good visibility on every score. With these conditions overhead and surrounding us, our final need was for good, level ground underfoot, and a great deal of it. Above all, we wanted ground with a minimum of people and houses on it, where rockets could rise, or crash, or even explode without wear and tear on neighbors' nerves. . . . The best answer to our needs were the high plains of east central New Mexico. A map of the region showed that the town of Roswell was situated near the center of the favorable area."

> "Every vision is a joke until the first man accomplishes it; once realized, it becomes commonplace."—Robert Goddard

12

New Home: New Mexico

Robert Goddard and his wife, Esther, drove into Roswell on July 25, 1930. Within days, they rented—and eventually bought—Mescalero Ranch, an agricultural estate with a comfortable adobe house and several outbuildings. For ten of the next twelve years, this would be their home. The only interruption would be in 1932–1934, when their funding was suspended because of the economic realities of the Great Depression as well as the death of Daniel Guggenheim and a restructuring of his family's charitable foundations.

Four assistants followed Goddard to Roswell: Larry Mansur and his brother Charles, Albert Kisk (Esther's brother), and Henry Sachs. As soon as they arrived, they set to work unloading the railroad freight car that was loaded with household goods, machinery, and rocket paraphernalia. A comfortable distance from the house, they had a 30-foot by 55-foot machine shop built. Two hundred feet farther away, they set up a 20-foot-tall tower for static tests of rocket engines (in which the engine was fired but restrained from moving).

Goddard designed a concrete base under the tower to divert the exhaust flame without kicking up a cloud of dust that would obscure the view of the rocket. He built a 3-foot-thick block of concrete encasing a 10-inch-diameter tube that bent from vertical to horizontal to channel exhaust blasting out of the combustion chamber in a direction away from the observation shelter, which was 55 feet away.

Among the equipment Goddard brought from Massachusetts was the 60-foot-tall launch tower from Aunt Effie's farm, although bringing a windmill frame to rural New Mexico was like bringing ice to Alaska. The launch tower was not destined for Mescalero Ranch, however. The headquarters location, only 2 miles from the center of Roswell, was too close to town for rocket launches. Goddard did not want to annoy or endanger the 11,000 local residents. Cattle rancher Cort Marley gave him permission to build a control shelter on his land, about 10 miles west of Mescalero Ranch. Marley's neighbor, Oscar White, let Goddard set up the launch tower on his adjacent property, in a hollow known as Eden Valley. Again, the crew built a concrete trough beneath the tower to deflect the fiery exhaust.

Becoming New Mexican

The Goddards found the people in Roswell to be friendly and helpful. Neither Marley nor White charged him anything for using their land, for example. They only asked that he close the gates when entering or leaving their properties.

Goddard and his wife soon felt more at home in the West than they did in the East. They loved the open spaces, and on weekends they could enjoy the forests, lakes, and cool air of the mountains only 70 miles west of town. The desert's low humidity was good for his health; he had lingering lung problems from a struggle with tuberculosis.

While still in Massachusetts, Goddard had taken up painting as a hobby. Now, captivated by the scenery in New Mexico, he sometimes wrote descriptions in his diary of cloud formations or the sky's colors at sunset. For example, an October 1934 entry included a simple sketch with this description: "Capitan [Mountain] pretty just after sunset. Ground dark brown. Mountain pale blue purple. Sky low down, to A [a point marked on the sketch], an apricot yellow (cadmium yellow and white) brightest at horizon. A to B, pale uncertain pink above B, dark gray blue, darker overhead. White Mountains also a pale blue purple, a very little paler than Capitan. All the angles on the mountains showed out sharply."

One of the Goddards' new friends was Harold Hurd, an Easterner who had come to Roswell for his health. Coincidentally, Hurd's son, Peter, was an accomplished painter who had been a student of Andrew Wyeth's father. Peter lived in San Patricio, but became a friend of Goddard's during visits with his parents in Roswell. After dinner on the evening they first met, Goddard, Harold, and Peter went outside and looked at the starry sky. "Dr. Goddard began pointing out the principal constellations and identifying by name their component stars—the ones of greater magnitude," Peter Hurd said in a 1970 Senate committee session honoring Goddard. "I recognized for the first time the glittering belt of Orion and learned that his left foot was the brilliant star Rigel. I saw the Great Bear with its pointer stars that aid in locating the Polar Star. So it was that on that evening began a continuing and to me most rewarding interest [in astronomy]."

In late October 1939, Goddard drove to Peter's home to sketch the landscape. "I don't recall what we talked about," Hurd told the Senate committee. "One basic thing I do remember: beneath a superficial shyness was a warm, friendly, unaffected personality. I think often of him, especially and poignantly when I delve into my books on astronomy."

Many Roswell-area residents liked the Goddards. Paul Horgan was a two-time Pulitzer Prize winning author who worked as the librarian at the New Mexico Military Institute in Roswell. Biographer David Clary quoted Horgan as saying, "[Goddard's] own dignity and his own nature were so appealing and winning that he was taken just as a fellow and liked very much, and Esther was popular and in a lot of things in town. . . . Aside from his work, I would say he was probably the most conventional human being I've ever known in my life. He was absolutely indistinguishable from a man who owned a lumber yard or a lawyer or a vice-president of a bank, highly conservative in everything except his wild interstellar obsession."

The townsfolk enjoyed the excitement of famous visitors like Lindbergh, but they respected the privacy Goddard wanted for his work. "Through the years, people would ask us where the Goddard [launch] tower was and we'd tell them stories," Marley's wife, May, told another Goddard biographer, Milton Lehman. "I told so many lies, I guess I'll never get to Heaven. We'd send them south of Roswell, usually. If anybody heard a rocket shoot and asked about the noise, we'd say it was the Indians up in the Capitan mountains, shooting cowboys, or we'd tell them it was the cowboys shooting the coyotes, or we'd say, if it was cloudy, that the noise was thunder—anything that popped into our heads."

Flying in the Face of Nature

Roswell's climate was what Goddard had sought, but Mother Nature could be cantankerous. Arriving in July, Goddard quickly noticed southeastern New Mexico's intense summer heat. His September 19, 1930, diary entry noted that the temperature in his office was 100 degrees. And heat was not the only challenge the Chihuahuan Desert offered. Another diary entry mentions that he "killed a tarantula who had made a hole near the base of the [launch] tower." Preparing for one launch, his assistants found black widow spiders and scorpions in one of the observation stations at the launch site—a 4-foot-deep concrete-lined pit with a heavy, iron-covered door that tilted up to shield the observer.

Larger critters could cause problems too. The crew learned to inspect the tower for obstructions after an unnoticed hawk's nest, constructed partly with several pounds of wire, ruined a launch. Sometimes, they got the rocket ready in the tower one day before launching it the next day. One of the assistants would stay at the launch site overnight to watch over the rocket and shoo away animals. In the quiet of the night, a mouse or prairie dog might decide to take up residence in the tower.

Even in the desert, there were rainstorms. Goddard's diary entry for August 8, 1940, read in part, "Went out with the men at 6:30 in morning. Got stuck in the mud near the Pine Lodge road for an hour and a half. Got things set up and E. [Esther] came with the Aldens at 1 p.m. Waited till 5 p.m. for the wind to go down. Tower struck by lightening last night with rocket in it. Tony [Ornelas, a Goddard crew member] sat in seat of truck 75 feet from tower, and saw the flame at the top of the upper guy [wire] on the left front, and a noise so loud he 'thought the rocket had exploded.'"

Mother Nature had created even greater havoc two years earlier. "At 3:15 [a.m.] a twister [tornado] had hit the tower, and made it a pile of rods and bent angles," Goddard wrote. "Truck and trailer 30 ft away OK. Charles [Mansur] was putting cover around lower part of rocket, and was therefore able to run from the tower. He heard a roar and saw a cloud of dust, and came to against one of the iron pipes supporting a guy wire. By this time it had passed." Goddard and the other assistants drove up at daybreak and saw the pile of rubble, but they did not see Charles. He was out of sight because he had just chased a rattlesnake under the trailer and shot it.

Flames in the Desert

By the time Goddard set up shop in New Mexico, it had been more than a year since his last rocket launch in Massachusetts. He had managed to conduct a few static tests on federal land since then, but he was anxious to get back to full activity. Now free of his teaching duties and able to work on rocketry full time, he made up his mind to conduct a launch before 1930 ended.

He just made it, with a launch on December 30. This rocket—still a skeletal structure—consisted of, from top to bottom, a parachute packed in a nose cone, a gasoline tank, a nitrogen tank, a liquid oxygen tank, and a combustion chamber and exhaust nozzle. The nitrogen was used to force the oxygen and gasoline into the combustion chamber. The nose cone and the three tanks were each capped with a sixty-degree cone, which could easily be formed by connecting the ends of a semi-circle of metal sheeting. At the bottom of the 11-foot-long rocket were four metal fins designed to provide stability during flight. Four sets of rollers mounted on the rocket's sides fit into rails on the tower to guide the vehicle directly upward.

After ignition, the rocket blasted skyward for 7 seconds before beginning its descent. It landed 1,000 feet away from the tower after reaching a height of 2,000 feet—ten times as high as any of Goddard's previous launches. The rocket reached a speed of 500 miles an hour. Certainly, technological changes he had made during the past year and a half were a major factor in the improvement. The launch's location gave it a small boost as well. Eden Valley's elevation was 3,000 feet higher than Aunt Effie's farm, so the air density was 9 percent lower.

The raw power of the rocket coupled with the possibility of an erratic flight path or an explosive failure led Goddard to devise a remote control system before attempting the next launch. At the control shelter, 1,000 feet from the tower, one member of the team peered through a telescope to read gauges at the tower and operated a sequence of switches that pressurized the combustion chamber, started the ignitor, and released the rocket restraints when full thrust was achieved. Another man timed the flight duration with a stopwatch. A third observer watched through binoculars to

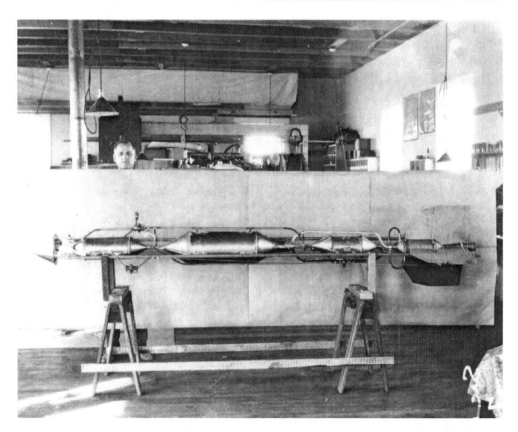

Goddard's first rocket launched in Roswell was the last to lack a streamlining shell. Henry Sachs, a machinist who came to Roswell with Goddard, held a backdrop for photographing the rocket.

note any rotation about the rocket's long axis. The fourth person was stationed 3,000 feet away from the tower with a recording telescope, which Goddard described this way: "Two pencils attached to this telescope gave a record of the altitude and azimuth [horizontal displacement], respectively, of the rocket, the records being made on a paper strip, moved at a constant speed by clockwork. The sights at the front and rear of the telescope, similar to those on a rifle, were used in following the rocket when the speed was high."

Goddard's wife, Esther, was also a vital member of the team. In order to document his work, she had learned to be a photographer, not only taking pictures but developing the films in her own darkroom. As components were redesigned and assembled into a new rocket, she photographed each part, the assembled vehicle, and the test setup. At the launch site, she was responsible for filming the entire flight with a movie camera and for putting out the scrub-brush fire touched off by the launch.

Robert Goddard reads pressure and lift gauges at the tower through a telescope. At his fingertips are three telegraph keys used to trigger launch events.

17

The second rocket flight in New Mexico took place on September 29, 1931. Not only did it use the new remote control launch system, but it marked a distinctive change in the rocket structure. To improve the vehicle's aerodynamic performance, the entire rocket was enclosed in a thin metal shell. The rocket was 10 feet long and 12 inches in diameter; with its tanks empty, it weighed 37 pounds—3½ pounds more than the previous version. It spent 9½ seconds in the air, reaching a maximum height of 180 feet. During the next four weeks, however, revised rockets reached heights of 1,700 feet and 1,330 feet.

The Complex Problems of Rocketry

Goddard knew that he had three essential problems to solve. The first, which he was currently focused on, was propulsion—getting the engine to develop and maintain enough thrust to push the rocket into the sky. Only after achieving that would it be appropriate to work on making the rocket fly in a smooth, stable path. Once that was accomplished, he could find ways to reduce the rocket's weight so it could be propelled higher.

Almost as important was a fourth problem: vehicle recovery. Developing a parachute system that would bring the rocket down to the ground without crashing would save the time and expense of repairing or replacing damaged components. Recovering undamaged components would also help with the analysis of how the components performed during flight. Furthermore, once rockets were reliable enough to carry experiments or passengers aloft, effective parachute operation would be necessary to recover instruments or occupants unharmed.

As a scientist, Goddard was well equipped to understand the physical principles needed to solve those problems. However, his compulsion to solve all of them sometimes led him astray from the principles of the scientific method. When he revised a rocket design after a static firing or a launch, he often changed several components before the next test. For example, he might refine the fuel injection system, adjust pressures in the propellant tanks, change the size and weight of the tanks, and modify the guidance system. As a result, the effect caused by any one change could not be isolated.

To the disconnect between scientific knowledge and the scientific method, Goddard added the tinkerings of a hands-on inventor. The son of a talented machinist, he

Over a period of years, a number of people from around the world volunteered to ride one of Goddard's rockets, even all the way to the moon. For example, one person wrote in April 1924 (even before the first successful liquid-fuel rocket launch), "I am willing to undertake the first trial journey in that machine into the Universe, and am ready for all kinds of events that may occur. I am thirty years of age, healthy, single, my parents are dead, no relatives, by birth Czechoslovak, my profession, reporter."

had always been fascinated by tools and mechanisms, and he seemed to think in three dimensions. When he needed to redesign a rocket component, he would sometimes sit in his shop, cut strips of metal from empty coffee cans, bend them into shape, and solder them together. Clary quoted Charles Mansur, one of Goddard's assistants, as recalling, "He'd cobble up some of the craziest looking monstrosities—nothing against him, he was a wonderful man—but he couldn't solder, he couldn't weld, he couldn't run a machine, but he did all of it, though. He'd get a big chunk of a thing set together and then it would all fall to pieces." Fortunately, Mansur and other team members had the necessary fabrication skills.

Propulsion

The motor, or combustion chamber, is the heart of a rocket. By the end of his first two years in New Mexico, Goddard had developed a reliable engine design. It was a cylinder, 5¾ inches in diameter and 11 inches long, made of sheet nickel. The top of the cylinder was closed with an insulated, shallow cone. The chamber bottom was a 120-degree cone, truncated to leave an opening about 1¾ inches in diameter. At this point was attached a 14-inch-long nozzle that flared to a diameter of 3½ inches at the far end. The combustion chamber proved to be so durable that it could be reused in repeated static tests and rocket launches.

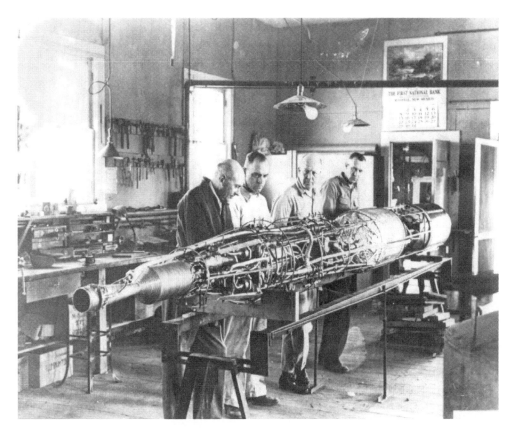

Goddard (left), Nils Ljungquist, Al Kisk, and Charles Mansur at work in Roswell. The rocket motor is at the left in the photo.

Figuring out how to inject the oxygen and gasoline in the correct proportions, pressures, and patterns was more difficult than settling on the overall chamber design. Goddard tried injecting them in opposing streams, injecting the gasoline through the center of the top and the oxygen from the sides of the chamber, injecting the oxygen from the top and the gasoline from the sides, and injecting both from multiple nozzles. A supplementary feature he used consistently was a "cooling curtain" of additional gasoline injected from tangential tubes on opposite sides of the chamber so that it flowed along the chamber wall before vaporizing. This kept the chamber wall cool enough to keep it from melting or burning through from the motor's sustained firing.

Moving the propellants from their tanks into the combustion chamber was another difficult challenge. For most of his work in New Mexico, Goddard used a tank of liquid nitrogen in the rocket. It would be vaporized in a controlled manner and fed into the gasoline and oxygen tanks to push the propellants into tubes leading to the motor. He designed the special valves, spray nozzles, gas pressure regulators, and propellant flow meters he needed to monitor and distribute the propellants and pressurizing gas.

Even the tanks were complex devices with custom-designed internal baffles to minimize sloshing of the liquids. Sometimes, wire or metal straps were wrapped around the tanks to strengthen them against the high pressure of their contents. At one point, Goddard even experimented with mounting the liquid nitrogen tank inside the liquid oxygen tank.

Igniting the propellant mixture in the combustion chamber was also part of the engine operation. The ignition system for Goddard's first successful rocket at Aunt Effie's farm was a blowtorch attached to the end of a pole. His assistant, Henry Sachs, standing next to the rocket's skeletal launch frame, had applied the torch to a combination of black powder and match heads. Later, Goddard used charges of explosive powder suspended in the combustion chamber and set off with a burning fuse. Eventually, he developed an electrical igniter similar to a spark plug.

The flight test of March 8, 1935, showed how powerful the rocket engines were. The motor fired for 12 seconds, producing a white flame only 8–12 inches long. The 75-pound, 14½- foot long rocket rose to a height of about 1,000 feet before it veered off course and traveled nearly horizontally. Just after the engine stopped firing, the parachute was released. But the rocket was traveling so fast, the ⅛-inch steel cable connecting the parachute to the rocket broke. The rocket finally landed about 9,000 feet from the launch tower. "From the distance covered during the last 6 seconds of the flight," Goddard wrote, "the top speed is estimated as about 700 miles per hour." This was the first liquid-fuel rocket to travel faster than the speed of sound.

Power Trips

For a while, Goddard tried to increase the rocket's power by making the combustion chamber a few inches longer and nearly doubling its diameter. After a series of unsuccessful tests, he tried something else. He built a rocket with four of the smaller motors clustered at the tail. One of the chambers burned through during the launch. Goddard thought this might have been caused by unequal flows of oxygen to the four chambers. He believed the multiple-engine rocket showed promise, but making it work would introduce a new set of problems to solve. He returned to the single, small motor model.

In an August 1937 launch, Goddard tried to boost the rocket's initial propulsion by catapulting it out of the tower. The rocket sat in a cradle of steel tubing. Rollers on the cradle, rather than the rocket itself, guided the vehicle straight up the tower. Two cables attached to opposite sides of the cradle rose to the top of the launch tower and passed through a system of pulleys that tripled the effect of two 40-pound concrete cylinders as they fell to the ground. In this way, some 240 pounds of push was added to the 200 pounds of thrust generated by the rocket's engine. "The rocket rose through the tower more rapidly than usual," Goddard wrote. "After the test the vertical ½-in. square tube supports of the cradle were found to have become somewhat bent." He concluded that the distortion of the cradle was "due either to inertia on stopping or to the fact that the 40-lb weights moved somewhat into the ground, thus pulling on the cradle after it had engaged the bumpers" that were designed to gradually reduce the speed of the cradle. The bending of the cradle apparently "caused the rocket to oscillate more strongly than usual."

On the Move

Rocket flights like the ones in March 1935 and August 1937 highlight the importance of stability in flight. If those rockets had not veered off their vertical paths, they would have reached much higher altitudes. In the first several Roswell launches, Goddard stabilized each rocket with metal fins at the tail, much like the feathers on an arrow. All of them veered significantly off their vertical paths.

Goddard came up with a new idea: movable vanes to steer the rocket back on course when it started to veer off. The system was based on a gyroscope, a device consisting of a disk attached to the center of a perpendicular rod, all of which is mounted inside two perpendicular rings that are attached to a frame in such a way that the disk and rod can spin and the rings can rotate within the frame. The important characteristic of a gyroscope is that while the disk is spinning rapidly, the rod will continue to point in the same direction even when the frame is tilted. Goddard's idea, then, was to mount a gyroscope in the rocket with the rod lined up with the rocket's long axis. If the rocket tilted away from vertical, the gyroscope would tilt with respect to the rocket. When that happened to a large enough degree, the gyroscope would make an electrical contact that would operate a system of levers to move the vanes in a direction that would steer the rocket back toward vertical.

In addition to four fixed vanes, the rocket now had four blast vanes that could move into the exhaust plume and four air vanes that could move out into the air stream surrounding the rocket. The vanes operated in pairs, with a blast vane moving inward at the same time the opposite air vane moved outward.

Goddard launched the first gyro-controlled rocket in April 1932. By this time, he had automated the launch control system so it required pressing only one switch. For this launch, starting the gyroscope spinning had to be added to the sequence. Four 5-pound weights were each attached to 4-foot lengths of fishline that had been wound so that dropping the weights unwound the lines and got the gyroscope spinning. When the control key was pressed, the weights dropped. This set off a Rube Goldberg-type operation that not only set the gyroscope spinning but also activated the entire launch sequence.

The rocket rose slowly out of the tower and reached a height of 135 feet before it turned downward and crashed into the ground under full thrust. Had the gyroscope idea fallen flat? Goddard rushed to the wreckage to feel the movable vanes. "The two vanes which, by entering the rocket blast, should have moved the rocket back to the vertical position were found to be warmer than the others," he wrote. He also noted that "the initial inclination of the rocket from the vertical appeared to take place more slowly than in preceding flight tests."

Goddard's first thought was that the blast vanes were too small to quickly correct the rocket's path. Later, he decided that they were so close to the nozzle opening that they partially blocked the exhaust flow and reduced the thrust. A noticeable bulge in the upper end of the combustion chamber after the test supported this conclusion.

Modifying the size and shape of the vanes became an ongoing effort, particularly since the rocket sizes kept changing. For example, during the next series of eight launch tests in 1934–1935, the rocket lengths varied between 13 feet 6 inches and 15 feet 3 inches, and their weights ranged from 58 to 85 pounds.

Ultimately, the system did help stabilize flights. In a March 1937 launch, for example, the rocket tilted into the wind shortly after leaving the tower but quickly returned to the vertical. Similar inclinations and corrections continued throughout the 22 seconds of propulsion, after which the corrections ceased and the rocket began to fall. Because of dust in the air and mechanical limitations of the tracking instruments, no accurate record could be made of its maximum altitude. However, the observers were able to watch the entire flight and estimated it reached a height between 8,000 and 9,000 feet. This was the highest launch Goddard ever achieved.

The rocket used for the next launch, in April 1937, was more than a foot longer and weighed 13 pounds more. Goddard redesigned the air vanes, nearly doubling them in size. The rocket quickly veered off course and flew horizontally under propulsion before crashing. The launch after that, in mid-May, used a rocket that was similar in length but 24 pounds lighter. Goddard again redesigned the air vanes. This time, the rocket effectively corrected its path throughout the 29 seconds of propulsion and even for a while after that. It reached a height of 3,250 feet.

Trying to correct the blast and air vane designs for each rocket was troublesome, so Goddard came up with another idea: a movable tailpiece. The combustion chamber was placed inside a conical tail section that could be swivelled. This would change the direction of the exhaust plume and alter the rocket's path.

The first launch using a movable tailpiece took place in late July 1937. According to Goddard, "the rocket rose rapidly with but little motion from side to side except at about half the maximum height reached, when it deviated about 30 deg[rees] on one side and immediately afterward to the same angle on the other side, thereafter proceeding vertically. After propulsion ceased, the rocket moved gradually toward the left and soon began to descend." The rocket reached a height of 2,055 feet.

A month later, in August 1937, Goddard launched another rocket with a movable tailpiece, this time assisted by a catapult. As described earlier, the catapult cradle hit the top of the tower and bent, tilting the rocket as it left the tower. The rocket corrected its path seven times, each time by a larger amount, until it stopped correcting at a height of 2,000 feet. The erratic flight caused the parachute to open too soon, and the rocket was torn apart.

Parachute Problems

Parachutes were standard parts of the Roswell rockets. The less damage a rocket suffered when landing, the more Goddard could learn about how each component had functioned during the launch and flight. Even badly damaged parts could be scrapped and the metal used to build new parts. Better yet, being able to reuse undamaged parts of the rocket saved time, effort, and money for the next test. At least once, he was even able to launch the same rocket twice.

Unfortunately, it was hard to get the parachutes to work effectively. One problem was deciding where to place the parachute. Goddard tried packing it in the nose cone or placing it below the gasoline tank, which was just beneath the nose cone. Once, he even put the parachute in a tin box that wrapped around the middle of the combustion chamber at the tail of the rocket.

Figuring out how and when to release the parachute was an even more difficult problem. Goddard tried several methods. First, he used a timer made from the inner works of a watch. At a preset number of seconds after launch, the mechanism ignited a small charge of black powder using the power from two small batteries. The explosion blew off a small section of the rocket's shell, and a spring forced the lid off the box containing the parachute. Predicting how long the rocket's engine would fire was difficult, though, particularly if the engine malfunctioned during flight.

For a couple of launches, he tried using an aluminum tube with a freely moving weight inside. The weight stayed at the bottom of the tube as long as the rocket was heading upward, but as the rocket turned downward, the weight moved to the other end of the tube and completed an electrical circuit that opened the parachute compartment. This apparently worked as planned in a March 1935 flight, releasing the parachute shortly after the exhaust flame stopped. However, the rocket was moving so fast that the steel cable connecting the parachute to the rocket broke and the parachute floated away uselessly.

Equipping rockets with a gyroscope for flight stabilization gave Goddard a better way to release parachutes. The same gyroscope that operated the movable steering vanes could be used to trigger parachute ejection. Steering corrections were activated when the rocket veered 5 or 10 degrees from the vertical, but a different mechanism released the parachute when the deviation was 90 or 130 degrees from vertical. This method was not foolproof, either. If an ineffective steering system let the rocket veer too far off course, the parachute could be released while the engine was still firing. In later tests, release was triggered by a decrease of pressure in either the oxygen tank or the nitrogen tank. Sometimes, release depended on a combination of low tank pressure and deviation from the vertical.

Lighter Tanks

The sponsors of Goddard's research were anxious for him to achieve greater flight heights, so he turned his attention to reducing the weight of the rockets as well as continuing to improve engines.

In 1938 James Doolittle, the renowned aviator who would later become a World War II hero, visited Goddard in Roswell. In his notes about the visit Doolittle wrote, "almost everything used in connection with these experiments, including all rocket parts, was fabricated in Dr. Goddard's shop." He also described the current rocket design, noting that "The cylindrical part of the tanks is used as rocket surface in order to save weight."

One way of reducing a rocket's weight was to use thinner metal for the propellant tanks. In late 1936, the gasoline and liquid oxygen tanks were made of sheet nickel. Because of the high pressure it had to contain, the nitrogen tank used slightly thicker sheet nickel and also had a 4-inch-wide band of much thicker nickel wrapped around its middle. The nitrogen tank weighed 22 pounds. The following spring, Goddard made the cylindrical nitrogen tank out of sheet nickel that was only half the previous thickness. He strengthened it by wrapping it with piano wire, both longitudinally and transversely. The wrapped tank weighed a little less than 11 pounds.

Wrapping the tank was tricky. The wire windings were spaced an average of a sixth of an inch apart and had to be done carefully to keep from deforming the tank. Goddard tested various metals and combinations of metal and wire thicknesses, many of which exploded under pressure. Once, after spending most of the day doing those tests, he received a phone call from a lawyer regarding a patent the rocket researcher held. "I had some trouble in hearing him," Goddard wrote in his diary. "His clients are Heinz and Kaufman, and I couldn't tell the first name until he said (yelled) 'pickles.' Afterward, E. [Esther] noticed cotton in my ears. I had stuffed them in because of the loud report, when the aluminum diaphragms I was testing burst, and I had had them in my ears while the man in San Francisco had been trying to talk with me. As a matter of fact, they were not cotton but pieces of rags."

The oxygen and gasoline tanks were not wire wrapped because they were smaller and more difficult to wrap successfully. Eventually, when a larger-capacity gasoline tank was used, the wrapping technique was used on it. The 20-inch-long tank was made of the same thin sheet nickel as the nitrogen tank and weighed 9 pounds. It was wrapped with a pound and a half of wire. "As an indication of the thinness of the tank wall, which withstood 460 psi when wired," Goddard wrote, "it may be mentioned that in the course of repairs made after the test the conical ends were cut off with an ordinary can opener."

Pump It Up

The other approach Goddard used to make the rockets lighter was to get rid of the nitrogen tank entirely. If he could pump the propellants into the combustion chamber, he would not need the pressurizing gas. Also, the propellant tanks would not be subjected to high pressures, so they could be made of thinner, lighter material. No commercially available pumps could handle the capacity and pressures the rockets needed, and none were capable of pumping liquid nitrogen. Over a period of years, Goddard invented, tested, and refined the pumps he needed.

The pumps Goddard designed and built functioned well when tested in the laboratory, delivering the pressures and flow rates he wanted. But when they were installed in rockets, none of the tests were successful, although the system did show promise.

In late 1941, Goddard put his pump development program on hold when he was hired by the Navy and the Army to apply his rocketry knowledge to aircraft modifications for World War II. After several months of preliminary work in Roswell, the Navy required him to move to Annapolis, Maryland, to continue the development. He was never to return to New Mexico. He died of throat cancer on August 10, 1945—four days before Japan announced its surrender, ending the war.

> After Goddard's death, Charles Mansur went to work at the White Sands Proving Ground and became the Chief of the Design and Preparation Section of its Propulsion Branch.

Resources

Clary, David. *Rocket Man: Robert H. Goddard and the Birth of the Space Age*. New York: Theia, 2003.

Goddard, Esther, and G. Edward Pendray, Eds. *The Papers of Robert H. Goddard, Including the Reports to the Smithsonian Institution and the Daniel and Florence Guggenheim Foundation*. New York: McGraw-Hill, 1970.

Goddard, Esther, and G. Edward Pendray, Eds. *Rocket Development: Liquid-Fuel Rocket Research, 1929–1941*. New York: Prentice-Hall, 1960.

Goddard, Robert H. "A Method of Reaching Extreme Altitudes," *Smithsonian Miscellaneous Collections*, vol. 71, no. 2, 1919. Accessed at http://www.clarku.edu/research/archives/pdf/ext_altitudes.pdf.

Lehman, Milton. *This High Man: The Life of Robert H. Goddard*. New York: Farrar, Straus and Company, 1963.

Chapter Two: Operation Paperclip and Beyond

Sometimes, it is simply time for a certain scientific or technological breakthrough to take place. Two or more people independently arrive at the same conclusion at about the same time. It happened with the discoveries of oxygen, sunspots, and evolution. It happened with the invention of telephones, color photography, and telescopes. It also happened with the invention of liquid-fueled rockets. Strangely, near-simultaneous developments in rocketry from different hemispheres converged in southeastern New Mexico.

Nearly a year after Robert Goddard moved to Roswell, Johannes Winkler became the second person to successfully launch a liquid-fueled rocket, in Germany. While working as an engineer for an aircraft company during the 1920s, he became fascinated with the idea of mankind traveling to the moon or to other planets. In 1927, he founded the Society for Spaceship Travel. The most prominent rocket scientists and engineers in Europe joined the Society, which grew to include more than 1,000 members by 1930. About that time, Winkler quit his day job to concentrate on rocket development with funding provided by an Austrian industrialist.

In March 1931, Winkler's 2-foot-long, 11-pound rocket rose to a height of 2,000 feet. He placed the combustion chamber at the head of the rocket, just as Goddard had done in his original design. However, in Winkler's version, tube-shaped tanks holding the liquid oxygen, liquid methane, and pressurized nitrogen gas were arranged in a triangle with an open center for the exhaust plume to pass through unimpeded.

Science historian Stephen Stigler investigated the phenomenon of simultaneous discovery and postulated a related conclusion: "No scientific discovery is named after its original discoverer." Ironically, this principle, known as Stigler's Law, had been developed previously by sociologist Robert K. Merton.

During the next year and a half, Winkler refined his design and built a much larger rocket. The first test launch in October 1932 was delayed several times, leaving the rocket exposed to the Baltic Sea's salty air for weeks. A corroded metal valve failed during the launch, and the rocket exploded at a height of only 10 feet. Either crestfallen or out of funding—or both—Winkler abandoned his research and went back to his old job at the aircraft company.

Other individuals and groups had also been working on rocket development. However, conducting rocket experiments was costly and funding was scarce. In 1932, several members of the Society for Spaceship Travel convinced the German Army to finance rocket research. Although those engineers and scientists would have to develop rockets as weapons, they expected to eventually apply the technology to space exploration vehicles. The following year, Hitler emerged as leader of Germany, and the development of rockets as instruments of war gained momentum. While Goddard built rockets in Roswell under a civilian grant, the Nazi government supported rocket development on a massive scale. In October 1942, the year after Goddard left New Mexico to work with the United States military, the Germans revolutionized rocketry by launching what would become known as the V-2.

The V-2 missile (a rocket used as a weapon) was huge and powerful: At liftoff, it weighed 27,000 pounds, including a 2,000-pound warhead. Its first successful launch reached an altitude of 62 miles, making it the first manmade object to travel beyond the Earth's atmosphere. By 1944, the Germans were building as many as thirty V-2s a day and firing them at Allied targets. More than 1,500 of them hit (or nearly hit) England, killing more than 2,500 people and causing extensive property damage.

White Sands

In 1944, with German rockets pummeling parts of Europe, officials in the US War Department realized it was time to do more with rockets than jet-assist airplane takeoffs. They decided to establish a facility for rocket development; the question was, where?

Just as Goddard had, they looked for a sparsely populated area with a large expanse of level terrain and a mild climate that would allow year-round experimentation. They found just what they wanted in New Mexico's Tularosa Basin, about 100 miles west of Roswell. The basin is bordered on the west by the Organ and San Andres Mountain ranges and on the east by the Sacramento Mountains—a geographic bonus providing weather protection and locations for elevated observation posts.

"The site, chosen after a survey of the entire United States, is flat, timberless, and sparsely populated, making it ideal for the firing of long-range rockets," photography expert Clyde Holliday explained in the October 1950 issue of *National Geographic*. "Cloudless skies prevail about 85 percent of the time, so that rockets can be followed throughout their flights by telescopes and cameras."

Another distinctive feature within the Tularosa Basin is a 275-square-mile expanse of pure-white sand, known—logically enough—as White Sands. No streams flow out of the basin, so accumulated precipitation forms temporary lakes. The water absorbs minerals from the ground, and white, sand-like gypsum is left as residue when the water evaporates. Constantly shifting dunes up to 60 feet tall comprise the world's largest gypsum dune field. President Hoover designated White Sands a national monument in 1933.

In 1942, because of compelling national interests following the United States' entry into World War II, over half of the Tularosa Basin was made available to the federal government for military training. A new rocket testing facility, called the White Sands Proving Ground (WSPG), was formally established in July 1945, initially encompassing 2,350 square miles. By 1952, it expanded to 3,200 square miles. Much of the land had belonged to the federal or state government, but 55 square miles of it had been rangeland owned by 129 cattle and sheep ranchers, who were ordered to leave their property, sometimes on very short notice.

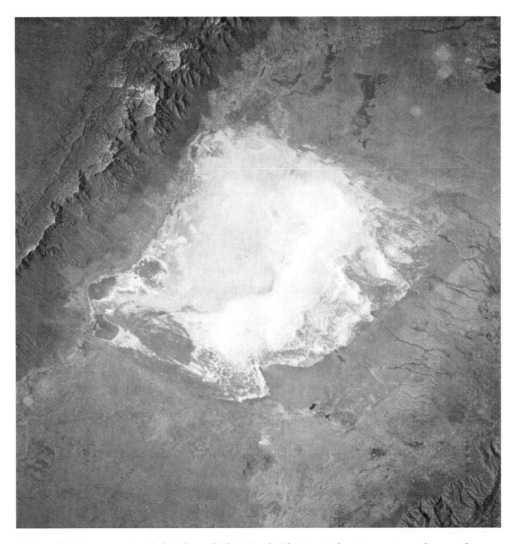

A Space Shuttle astronaut took this photo of White Sands. The San Andres Mountains angle across the upper left, and the Sacramento Mountains appear in the lower right.

The transition was not a smooth one. The government sent the ranchers lease payments for years, telling them that they needed to stay off their land only temporarily. Relocating their herds—and their families—imposed a hardship on the ranchers. In addition to buying or renting new grazing lands, they had to continue paying property taxes on the land they could no longer access. Years later, the government condemned the properties and offered a minimal purchase price. Some of the families refused the offer. Government lease payments ended in 1982, with some ranchers feeling they had not received full compensation for the permanent loss of their homes and rangeland. Land owners filed law suits and staged at least two armed confrontations. Ultimately, the government prevailed.

Colonel Harold Turner, WSPG's inaugural commander, drove to New Mexico from Washington, DC, passing through Roswell and across the Tularosa Basin, arriving near San Agustín Pass in early July 1945. "Having come from the East and having been raised primarily as an Easterner, the view was, to put it mildly, startling," Turner wrote in a 2008 article. He got out of his car to look around. "As we walked south on this [old ranch] road, on my right hand, I could see the majestic Organ Mountains. At that time I did not know the name. And on my left, the broad expanse of the desert known as the Tularosa Basin. At this time, this was in fact a broad expanse of mesquite humps as far as the eye could see, sand and apparently nothing but sand."

After spending the night in Las Cruces, New Mexico—25 miles across the Organ Mountains from White Sands—Turner set to work turning WSPG from an idea to a physical reality. To hurry construction along, he scrounged temporary buildings from Sandia Base in Albuquerque and moved them to White Sands. While they were being moved, he chose the spot for WSPG's main base, about 30 miles southeast of the gypsum dunes.

Turner had been at White Sands only a week when he was startled by the sound of a huge explosion off in the distance. Word came that a large ammunition dump had blown up. He did not learn until later that what had actually blown up was the world's first atom bomb. The Trinity Site, 90 miles north of his camp, would eventually become part of the New Mexico rocket testing facility, which would be renamed the White Sands Missile Range (WSMR, pronounced *Whiz-mer*) in 1958.

Von Braun's fear of execution was well founded. In March 1944, he had been arrested by the Gestapo and held for about two weeks. During his interrogations, he learned that he faced several charges, including an assertion that he planned to sabotage Germany's rocket program by developing a spaceship instead of a ballistic missile. In support of that charge, the Gestapo cited a short story, "Lunetta" (Little Moon, actually an orbiting space station) that he had written in high school. He was released from custody only because of his importance to the V-2 project.

V-2s Come to White Sands

As the WSPG base was taking shape, Germany was falling apart. In February 1945, Allied forces pressed into Germany from several directions. It didn't take a rocket scientist to realize that the war—and the Third Reich—would end soon. Indeed, the scientists and engineers working on the V-2 program realized it and began to consider their options. They feared being shot by the SS (Nazi Protection Squad) to keep Germany's conquerors from gaining their rocketry knowledge. If they were not executed by their own government, they would be at the mercy of their captors—Soviets, French, British, or Americans.

Wernher von Braun, one of the key inventors of the V-2, organized a large group of missile personnel to prepare for their destiny. During the first half of March, they hid rocket components, equipment, and important paper documents in every factory, mine, and warehouse they could find near the V-2 factory. As time was running out, von Braun led them out of the German-controlled territory so they could plan their surrender. Many of them had ties to the Nazi regime—von Braun, himself, was a captain in the SS, although it was more for expedience than ideology. He later told US officials that the only time he ever made use of his SS rank was to get his group through a German roadblock when they were trying to escape the crumbling Reich.

With the large group in a place of relative safety, von Braun and six other prominent members set out to find the US Army and arrange a surrender. "Some 150 of

Charles Stewart, left, accepts the surrender of Wernher von Braun, center, and his companions. Von Braun had broken his left arm in a car accident several weeks before.

the top German rocket personnel . . . wished to join the Americans to continue their work in rocket development," Charles Stewart, an Army intelligence agent with the American unit they found, told the authors of *Space Travel: A History*. "They had selected the Americans, as they were favorably disposed to this country generally and also because this country was the one most able to provide the resources required for interplanetary travel."

The group was well received. The self-assured von Braun told a reporter in 1950, "I didn't expect to be treated as [a war criminal]. No, it all made sense. The V-2 was something we had and you didn't have. Naturally, you wanted to know all about it."

Not surprisingly, the US government accepted the group's surrender and their offer of the large amounts of V-2 material they had hidden. Their relocation to the United States was given the code name "Operation Paperclip." The Army had to move quickly, however. Clearly, the area around the rocket factory was going to be claimed by the Soviet Union. Over a ten-day period, ending the day before Soviet forces were expected to arrive, US soldiers and German factory workers loaded a total of 341 freight train cars with hardware and fourteen tons of documents, and sent them off to Antwerp, Belgium. From there, the material was loaded onto Army cargo vessels, shipped to New Orleans, repackaged into train cars, and sent to Las Cruces.

Three hundred train cars of V-2 material arrived in mid-August 1945. It took every train track siding from El Paso, Texas, to Belen, New Mexico—a distance of 210 miles—to store the freight cars until they could be unloaded. Over a thirty-day period, the material was unloaded, placed on trucks, and driven over San Agustín Pass to WSPG. Every flatbed truck in the area was pressed into service. With limited warehouse space at the new base camp, many of the sturdier components were stored outside. Some of them were already corroded with rust from their seaside location in Germany. Over time, sand abrasion and infiltration damaged them further.

There were no complete V-2 rockets, but there were many components, including some 200 engines, 180 sets of propellant tanks, and 90 tail assemblies. There were also assorted mechanical and electrical components in varying numbers. In fact, there were only enough original parts to build two complete rockets. The Army had to find American companies who could duplicate and manufacture many replacement parts. They needed to design some new parts also. For instance, some German components could not connect to or communicate with American launch facilities. And new nose cones capable of carrying scientific experiments were needed to replace the original warheads.

Germans in New Mexico

The German rocketeers who came to the United States under Operation Paperclip were dispersed to several sites around the country to work on various projects. Fifty-five of them, including von Braun, came to Fort Bliss, which straddles the New Mexico/Texas border near El Paso, to support the V-2 program at White Sands. They lived at the army base and were bused to WSPG when necessary.

"When I was first informed that I was going to be sent to El Paso I was intensely curious to see it," von Braun wrote in a 1952 magazine article. "I knew it was part of the great American 'Wild West.'" Rather than carousing with cowboys, however, he found himself restricted to the base or closely supervised off base. The Germans' presence in the United States was kept secret for more than a year. The ones assigned to White Sands called themselves "prisoners of peace."

Von Braun and his compatriots developed friendships with the other engineers who worked at WSPG. On one occasion, they invited some of these Americans to a party at Fort Bliss, complete with delicious German cuisine and home-brewed beer. "The favorite subject for discussion that evening was the fuel used in the V-2, which was Ethyl alcohol, or Ethanol," Thoral Gilland, a Rocketdyne engineer who frequently worked at White Sands, recalled. "Our hosts explained that, aside from a few small fields in Romania, Germany had no petroleum reserves. Alcohol was the most popular fuel for their transportation needs, as well as their war machines. Early in the war, German agronomists developed a huge hybrid potato that was the favorite feedstock in Germany's alcohol distilleries, which kept the wheels rolling in Germany, literally." During the late stages of the war, when food supplies dwindled in Germany, workers at the V-2 development station comforted themselves by consuming alcohol intended to be rocket fuel.

One of the first tasks assigned to von Braun and his team was to teach the Americans how to assemble and launch a V-2 missile. The Army had contracted with General Electric Corporation to develop a series of American rockets. Known as Project Hermes, that program included assembling and testing the German missiles. The deteriorated condition of some of the V-2 components and the scarcity of certain parts made this difficult. Assembling the first complete rocket took months. After a successful static test of the engine, the rocket was ready for launch at WSPG on April 16, 1946.

As the countdown progressed, a crowd of technicians, scientists, and other observers gathered in the 20-foot by 40-foot launch control room in a protective blockhouse. It was a concrete hulk of a building, with 10-foot-thick walls and a pyramid-shaped roof up to 27 feet thick. It had been designed to withstand the explosion of a fully fueled V-2 at a distance of 100 feet or the impact of a V-2 falling on it at 2,000 miles an hour. Naturally, window space was minimal. The north wall of the blockhouse, facing the launch pad, had only two 6-inch-high, 3-foot-wide windows filled with thick glass.

As soon as the rocket safely zoomed skyward, everyone wanted to see its flight. They flung open the blockhouse door and ran outside. The experts tracking the rocket's path quickly realized that it was behaving erratically and aborted the flight. At that point, the rocket began to fall back toward the ground. The people who had just run outside turned to rush back into the blockhouse, but ran into the other observers who were pushing out of the building. While everyone struggled in this human traffic jam of confusion and curiosity, the rocket smashed into the ground 5 miles away.

The next launch, a month later, flew better. In its nose cone, it carried a scientific payload—a 35-mm camera and instruments for measuring cosmic radiation (fast-moving subatomic particles and electromagnetic waves generated by the sun and stars)—to a height of 70 miles. The researchers got no data or photographs, though, because the rocket landed head first, obliterating the nose cone and its contents. At von Braun's suggestion, some subsequent test flights carried the payload in the rocket's tail section, which was more likely to survive the landing.

A Journalist Reports

As necessary parts became available and the Americans became more skilled at assembling the rockets, V-2 tests became more frequent, averaging one every two weeks. A March 1947 *New Mexico* magazine article by Orren Beaty illustrates the flavor of the static tests and launches. For example, he described a static test: "There the 46-foot, 14.5-ton V-2 has been tested, with spectators permitted close enough to feel the full impact of the ear-punishing roar of the missile's rocket engine, expending 10 tons of liquid oxygen and ethyl alcohol in less than 70 seconds."

The actual launches were even more dramatic. The assembled rocket was trucked to the launch pad and raised into vertical position. "From then until the star flare is fired from the control blockhouse 500 feet away, signifying the launching is two minutes away, everyone concerned works at top speed," Beaty wrote. "Ethyl alcohol, 5,000 gallons of it, must be loaded into one set of fuel tanks; another 5,000 gallons of liquid oxygen . . . must be transferred to other fuel tanks in the V-2. Scientific instruments have to be set in place, checked and tested. Igniters are inserted. . . . All wiring is tested."

Completing those pre-launch preparations took several hours. During that time, the rocket stood upright, only partially enclosed by a 65-foot-tall, bright red, metal scaffold. For a daytime launch in the summer, the desert heat could be intense. Once, when the rocket's nose cone contained a camera, a canvas shade was used to keep the film from melting before launch.

When the rocket was finally ready, the launch sequence began. Inside the rocket, steam turbine engines fueled by a chemical reaction between hydrogen peroxide and potassium permanganate began pumping the propellants into the combustion chamber. The igniter was fired.

During 1947, control of the V-2 modifications and launches of scientific experiments at White Sands transitioned from the Germans to the Americans. Frustrated by the slow progress of the US space program, von Braun tried to stir up public enthusiasm by writing a science fiction novel, Mars Project. Although a brilliant engineer, he was not an effective novelist, and his manuscript was consistently rejected by publishers.

"There is a puff of smoke and fire at the base of the rocket. The tips of the fins, on which the huge missile is resting, are obscured by dust and smoke," wrote Beaty. As the 5,000-degree exhaust plume grew, the rocket started to rise. "Now a much louder roar can be heard. It pounds at the ears, while the rocket . . . accelerates rapidly. . . . It's easier to watch the rest of the flight lying down, as the rocket rises so rapidly it seems to be directly above the observer, and the neck strain is terrific."

Several miles into the sky, the rocket created a vapor trail. "It . . . gives the impression that the missile is engaging in acrobatics, as winds of varying directions and intensities at the different levels give the vapor trail a corkscrew effect." Beaty wrote that as the rocket rose even higher, the vapor trail stopped but he could still see a tiny dot of flaming exhaust. When the engine stopped firing, the rocket disappeared from view. "The roar of the rocket can be heard for several seconds longer, then the echo from nearby mountains lingers a moment more."

The scene Beaty witnessed was one of sixty-seven attempted V-2 launches at White Sands between April 1946 and October 1951. Two-thirds of them were successful.

Keeping an Eye on the Rockets

The V-2 launches at WSPG served two purposes. One was carrying instruments and experiments to investigate the Earth's upper atmosphere. Sensors measured temperature, air pressure, or the presence of certain gases as the rocket climbed to a height of 100 miles or more. For example, as Holliday wrote in *National Geographic*, "Samples of air at various altitudes have been captured in steel bottles strong enough to survive the rocket's fall. The samples disprove a long-held belief that helium and hydrogen, being lighter than oxygen and nitrogen, would rise higher in the Earth's atmosphere and concentrate at the upper levels. The bottles' contents showed that the proportions of these four gases in the air is no different up to 45 miles aloft than at sea level."

Scientists also sent various living organisms aloft to determine whether they would be harmed by cosmic radiation. A July 1946 flight, for example, carried corn seeds and fruit flies. After the rocket landed, the corn seeds were planted to see if they would grow normally, and the fruit flies were observed to see if their mating habits or offspring would be normal.

"The Holloman laboratory also took charge of mating fruit flies exposed to cosmic radiation in an effort to investigate genetic effects. This branch of study was, to cite Major Simons, of 'primary interest to pilots in terms of morale.' Yet in all the experiments performed 'in house' by Holloman aeromedical scientists . . . the effects of cosmic radiation have so far proven either negative or simply inconclusive." *History of Research in Space Biology and Biodynamics*, Air Force Missile Development Center, 1958

The other purpose of the launches was to understand the behavior of the rockets themselves. This would enable engineers to improve the accuracy and reliability of the V-2 design and, ultimately, apply the knowledge to designing new types of rockets. Sophisticated tracking telescopes were needed to observe and record the rockets at the speeds and distances they flew.

Clyde Tombaugh came to White Sands in August 1946 as chief of optical measurements. He thought of himself primarily as an astronomer, and in fact, he was famous for his 1930 discovery of this solar system's ninth planet, Pluto. To refine his searches of the sky, he had been building telescopes for twenty years, even hand-grinding his own lenses and mirrors. Understanding optical instruments and knowing how to track and photograph moving objects were the skills WSPG needed. At first, Tombaugh was not sure he was interested in switching from star watching to rocket watching. When he visited White Sands, though, he became excited about the project's challenges. Years later, he told his biographer, David Levy, the reason he made the move: "If it's telescopes, I'm there."

Tombaugh's visit to White Sands may also have revived a childhood fantasy. "I used to think about how nice it would be to visit the planets," he said in a 1991 interview. "Of course, I didn't expect to see in my lifetime what has happened. I knew it would happen some day, but it came along faster than I at first thought."

He also said, "We know now from sampling with big telescopes, that the number of stars in the skies is 10 to the 21st power. Now, that doesn't mean anything until I tell you that the number of grains of sand in all of the Earth's ocean beaches is only 10 to the 19th power. So there are a hundred stars to every grain of sand in all the ocean's beaches. They're not all sterile. How could they be? . . . We are not the center of the universe. We are not all that important. And we're not alone. That's my perspective."

During his nine years at WSPG, Tombaugh developed the optical systems for several telescopic photography devices. He mounted his camera-equipped tracking telescopes on surplus .50 caliber machine gun turrets. Because the rockets moved so quickly, two operators were needed for each tracking telescope—one controlling the device's horizontal movement and the other controlling the vertical movement. The telescopes were powerful enough to get sharp images of the rocket at a height of 100 miles. The films revealed details that could not be obtained any other way.

In 2006, the International Astronomical Union defined *planet* in a way that excluded Pluto. It is now considered a *dwarf planet*. The New Mexico state legislature promptly passed a memorial addressing the issue. Citing the state's long and extensive history of astronomical achievements as well as its recognition of long-time resident Clyde Tombaugh, the memorial resolved that "as Pluto passes overhead through New Mexico's excellent night skies, it be declared a planet."

In one case, the images showed that the rocket started to wobble just as the fuel supply ran out. Analysts could see that at that moment, the small amount of remaining fuel burned toward one side of the nozzle rather than uniformly around it. In future flights, engineers installed a device that would ignite all of the remaining fuel in the pipes a moment before it ran out. This solved the lopsided burnout problem.

One of the tracking telescopes was equipped with a diffraction grating (an optical device that splits light into several separate-color beams) so it could record the light spectrum of a rocket's exhaust plume. To get the best results, this device had to be located as close as possible to the launch pad. Twice, Tombaugh and his associate operated the tracking telescope as close as 300 yards from the pad.

"We put cotton in our ears, and it was an experience I'll never forget," Tombaugh told Levy. "The vibration just shook us like mice. An accident, an explosion at the launcher would have killed us. After two launches we figured we had used up our luck and we stopped."

Each rocket launch was simultaneously recorded on several tracking telescopes and other still and motion picture cameras. Finding the best locations for the tracking telescopes was one of Tombaugh's tasks. One site he chose was 200 miles away on Mule Peak in the Sacramento Mountains. Another was on San Andres Peak, on the opposite side of the Tularosa Basin. It was hard to get to, but it provided what Tombaugh described as "such a superb view." Observers used it for a while, but eventually stopped—partly because of the access difficulty and partly because a lightening bolt nearly struck one of the observers.

Students and Broomstick Scientists

Rockets were fired and data was collected at White Sands, but much of the data was analyzed in Las Cruces. In 1946, New Mexico A&M College (now New Mexico State University) established the Physical Science Laboratory (PSL) to process raw data from the rocket tests.

The development of digital computers was in its infancy at that time. The word *computer* generally meant a person who performed computations. At PSL, the computers were college students who worked part time processing the WSPG data. Albert Rosenfeld wrote an article for the March 1949 issue of *New Mexico Magazine* describing what he saw on a visit to PSL: "With rows and rows of electric calculating machines at their command, computers write down long lists of figures under columnar headings involving complex trigonometric formulae. Film readers record mysterious numbers centered at linear intersections; part-time student workers laboriously measure spaces between uneven wavy lines; researchers watch green electronic vibrations flicker across a black background [on an oscilloscope]. Typists pound out voluminous reports dealing with a bewildering complexity of intricate mathematical expressions and full of technician terminology."

The electric calculating machines Rosenfeld mentioned were mechanical devices that could add and subtract, and (using repetitive additions and subtractions) multiply and divide. The students filled in data tables and drew graphs by hand. "These unpublicized projects go on quietly and unsensationally, like the rather dull-seeming daily routine of any business office; yet, the purposeful, methodical activity is highly significant," Rosenfeld wrote. "Without it the rocket's roar across the desert sands would still be something out of Buck Rogers instead of an actuality."

While the PSL's student employees were processing data, other New Mexico A&M students were working at WSPG on rocket development and tracking. In 1952, the College instituted the Cooperative Student-Trainee Program. To be accepted into the program, a high school graduate had to have good grades, withstand a security clearance investigation, and score highly on a scientific aptitude test. Once in the program, the student would alternately attend classes for six months and work as a full-time employee of WSPG for six months. Coop students took five years to earn a Bachelor's degree, but the wages they earned working at the proving ground enabled them to pay for their education.

As the Korean War began, the government transferred von Braun and the other German rocketeers to Huntsville, Alabama, where they worked on developing a new rocket, the Redstone. At White Sands, enough parts were still available to assemble a few more V-2s, but Americans were also designing new rockets and missiles. The Asian war provided an unusual opportunity for increasing the technical personnel at WSPG. The Army was drafting young men and recalling veterans to active duty. Those who had university degrees—especially in engineering, mathematics, or physics—were assigned to perform their service at a few missile development facilities, including Huntsville and White Sands.

Those assigned to WSPG comprised the V-2 Section of the First Ordnance Guided Missile Support Battalion. An Army Captain at the base resented how rapidly these men were promoted on the basis of their civilian experience. Once, he told them he was no more impressed with them than he would be by a bunch of broomsticks. The men turned the attempted insult into a badge of honor, thereafter calling themselves the Broomstick Scientists.

Even though assembling and launching rockets was serious and sometimes dangerous, the Broomstick Scientists found ways to inject humor into their work. For example, when they had finished building their first V-2 rocket and moved it to the launch pad for a test firing, they hung a sign on the assembly building that read "OUT TO LAUNCH." In the blockhouse, they relabeled some of the warning lights on the launch control panel. One that indicated a malfunction requiring a pause in the countdown was labeled "TILT." The one that indicated a malfunction during engine ignition was labeled "AW SHIT!"

"There was an incident where we swore someone was firing a .30 caliber rifle every few minutes and in the vicinity of the rocket," Arnie Crouch, one of the Broomsticks, wrote in a 1998 article. After clearing the area, soldiers searched for the shooter. Everyone was surprised when they finally found the sound's source. Crouch explained, "Static electricity generated by the dry desert wind blowing across one dipole antenna mounted on the gantry's top . . . created an arc of 4 to 5 inches and a crack like a rifle shot when the ungrounded loose end of the antenna cable would be blown near the frame of the gantry. It was soon secured and grounded. Work resumed."

Not only did the Germans leave WSPG in 1950, but General Electric was nearing the end of its V-2 experiments. The Broomstick Scientists were expected to learn how to assemble and launch V-2s from the dwindling supply of parts, and then to work on other Army missile projects. After training with the General Electric staff for several months, they were ready for their final exam. They would build a V-2 by themselves and launch it, with the objective of setting a new altitude record. Knowing that reducing the rocket's weight would help extend its flight, they modified the design and were able to leave off several parts.

The launch was set for August 22, 1951. The missile sat on the launch pad, surrounded by the movable gantry. "The rocket's fuel is ignited by a cheap pinwheel type of pyrotechnic," Crouch wrote. "This is suspended by kite sticks in the motor's combustion chamber, about 3 feet up in the [combustion chamber]. The only way to install it is to literally get your head and shoulders up in the throat of the motor." When it was time to fire the igniter, the pinwheel unexpectedly flew out of the engine and skittered across the ground. "The good part was we could put the countdown on hold. . . . The bad part was, someone had to literally crawl up in the motor of the fully fueled rocket to install another igniter," Crouch wrote. One of the crew volunteered for the dangerous task, and installed a new pinwheel. The countdown resumed and the rocket performed flawlessly, setting a single-stage rocket altitude record of 132.6 miles.

Not the Only Game in Town

With their darkly glamorous history and awesome power, the V-2s attracted everyone's attention. But they were not the only rockets tested at White Sands, nor were they the first.

The Army first launched its WAC Corporal rocket on October 11, 1945, at WSPG. Not designed to be a weapon, it carried instruments for gathering information about the upper atmosphere. Its inaugural flight reached a height of 44 miles—twice what was expected. Von Braun's team arrived about two months later, and WAC Corporal launches continued during the V-2 program.

Ultimately, in Project Bumper, the two missiles became a tag team. In May 1948, a WAC Corporal mounted on the nose of a V-2 became the world's first two-stage liquid-propelled rocket. Several more Bumper rockets were launched at White Sands during the following two years, with the most memorable taking place in February 1949. The V-2 fired first, reaching a speed of 2,600 miles per hour before separating from the WAC Corporal, which then fired. It reached a speed of 5,100 miles per hour at burnout, coasting to a height of 244 miles before falling back to Earth. In 1950, the project moved to Florida and a Bumper became the first rocket to be fired from Cape Canaveral.

Over the decades, several rocket types have blasted off from White Sands, either carrying experiments or being tested themselves. Some have accomplished historic

A WAC Corporal missile looks pencil thin on top of a V-2 in this Bumper rocket. The diagonal paint markings on the V-2 helped film analysts watch its rotations.

In order for a test to be successful, all parts of the rocket, as well as any experimental payload, had to be recovered for analysis. Pieces could be scattered over a large area, and they were often buried in the sand by the impact of landing. Searching for them was time consuming and not always fruitful. During the 1960s, a crew of canines solved the problem. Before launching a rocket, technicians would spray important parts with shark liver oil. Humans could not smell the oil, but the dogs could sense it hundreds of feet away. The eight Missile Dogs found 96 percent of the objects they searched for. Cynthia Guzevich, a co-owner of the dogs, outfitted them with terrycloth jackets. In hot weather, the dogs' handlers put ice cubes in several pockets in the jackets to keep the animals comfortable. The best sleuth team was a German Shorthair named Count and a Weimaraner named Dingo.

firsts. The Redstone, which von Braun helped develop in Alabama, underwent testing at White Sands. In 1961, a Redstone launched from Cape Canaveral sent the first American astronaut into space, carrying Alan Shepard on a suborbital ride to an altitude of 116 miles. Later that year, another Redstone carried Virgil (Gus) Grissom 118 miles high in the Mercury Program's other manned, suborbital flight.

In addition to testing the rockets themselves, WSMR sometimes used rockets to test other objects. Two examples come from the Apollo Program, which sought to take astronauts to the moon. One of the objects tested at White Sands was the launch escape system for the Jupiter V rocket that would send the men into space. The system would enable the astronauts to separate from the rocket if it failed during launch. The Jupiter rockets were expensive, so a less costly (but also less aesthetic) stand-in was used for testing the escape system. It was a chunky vehicle called Little Joe II (its predecessor, Little Joe, had been used to test the Mercury Program's launch escape system). A Little Joe II standing now on a hillside beside the New Mexico Museum of Space History in Alamogordo is visible from 35 miles away. The largest rocket ever launched in New Mexico, it stands 87 feet tall.

When an Apollo flight reached the moon, the command module would go into orbit. The Lunar Excursion Module (LEM) would take the astronauts from their orbiting capsule to the moon's surface and then, after their exploration, lift them from the surface to rendezvous with the command module. Before the first Apollo mission, the LEM was tested at White Sands to ensure that its ascent, descent, and direction control systems worked properly. Two facilities constructed at WSMR for static tests of the LEM could simulate the lunar atmosphere by drastically reducing the air pressure around the vehicle.

More recently, all of the Space Shuttle astronauts have trained extensively at White Sands, practicing landings at the facility's 6.6-mile-long runway. In 1982, the Shuttle Columbia actually landed at WSMR when heavy rains literally flooded the primary destination, Edwards Air Force Base in California. The weather was not ideal at White Sands, either, as strong winds whipped up a dust storm that delayed the landing for a day before slowing down to tolerable levels. "The weather was atrocious," said Alamogordo resident Don Larsen. "White Sands was airborne." An estimated ten thousand people came to see the Shuttle land. Following the announcement of the overnight delay, the excited observers turned the evening into a tailgate party and then slept in their vehicles. The next morning, they cheered and clapped as the Shuttle landed perfectly.

Spacecraft work at White Sands goes beyond rocket performance tests. For example, before their missions, both the LEM and, later, the unmanned Viking Surface Sampler were prepared in a clean room at WSMR so they would not contaminate the soil when they landed on the moon and Mars, respectively.

WSMR conducts tests for commercial customers as well as government space programs. According to the facility's web site, "The mission of White Sands Missile Range begins with a customer, a service developer, or another federal agency, which is ready to find out if engineers and scientists have built something which will perform according to job specifications. . . . We shake, rattle and roll the product, roast it, freeze it, subject it to nuclear radiation, dip it in salt water and roll it in the mud. We test its paint, bend its frame and find out what effect its propulsion material has on flora and fauna. In the end, if it's a missile, we fire it, record its performance and bring back the pieces for post mortem examination."

Missile Mishaps

The raw power of rocket engines makes for spectacular flights. When tests fail, that power can be horrendous. The following examples illustrate what can happen on—or off—the White Sands Missile Range.

Just after four o'clock in the afternoon of May 15, 1947, a V-2 left the launch pad and immediately experienced steering problems. It soared 76 miles into the sky before exploding. The largest surviving section fell to Earth 3 miles outside of Alamogordo, but smaller chunks landed at several places in the city. "One woman called up the Army to 'get that thing out of my backyard,'" *Time* magazine reported in an October 10, 1949, article about WSPG's safety program.

Exactly two weeks later, a Hermes II rocket was launched. It was a two-stage rocket that used a modified V-2 as the first stage. For this, its first test flight, it was supposed to head north over WSPG. Its gyroscopic guidance system malfunctioned, and it headed south instead. The flight safety officer realized what was happening but decided not to abort the flight, fearing that the fuel-filled rocket would fall and explode in a populated area. The rocket flew for 5 minutes before it ran out of fuel, passing over El Paso and crashing 3½ miles south of the Ciudad Juarez, Mexico, business district.

"The violent blast, which shook virtually every building in both El Paso and Juarez, startled citizens of the two cities, who swamped newspaper offices, police headquarters and radio stations with anxious telephone inquiries," reported the *El Paso Times*. "The terrific impact of the rocket, which contained only telemetering equipment, scooped out a perfectly rounded crater, about 50 feet in diameter and 24 feet deep. . . . The site is half a mile from Buena Vista airport, where 13 planes were shaken by the blast, and a mile and a half from an oil plant. Many Juarez citizens at first believed the oil plant had exploded."

"A few hours after the wayward missile landed, the US Army showed up and found that enterprising Mexicans were selling any old piece of scrap metal they could find and claiming it was V-2 debris," wrote Wayne Mattson in the June 2001 issue of *New Mexico Space Journal*. "The United States ultimately apologized to Mexico for the incident and paid for all damages incurred." The crashed rocket was widely reported to be a V-2, and the Army let the misperception stand because Hermes II was a secret project.

In June 1951, a V-2 hit closer to home. As thrust was building toward liftoff, the rocket exploded on the launch pad. Arnie Crouch was in the blockhouse at the time, and in a 1998 article, he described what happened: "Imagine your reaction at having 4½ tons of alcohol and 5½ tons of liquid oxygen exploding just 50 yards away from you. The blockhouse felt like it was bouncing up and down, even though it had 8- to 10-foot-thick concrete walls and a 27-foot deep pyramidal concrete roof. We knew we were safe from the explosion; however, our immediate concern was the possibility of being fried alive if the burning liquid oxygen and alcohol mixture from the missile's ruptured fuel tanks should flow [through] the tunnel used to carry control wiring from the blockhouse to the missile." Fortunately, the blockhouse designers had foreseen this possibility and included some vertical segments in the cable tunnel to prevent liquid from flowing into the shelter.

A June 1952 launch was not supposed to happen. A Viking rocket engine developed by the US Navy was tightly secured to a static test stand. The engine would be held in place during the firing so instruments could measure the thrust it generated. However, the connections failed and the engine flew upward. It was not even a complete missile, so it was not aerodynamically stable and had no guidance system. The observers stayed in the safety of the blockhouse until they felt a distinctive thud. The engine landed 5 miles away. Jim Eckles, WSMR's public relations manager, wrote in 2006, "To this day, the Navy holds the White Sands' (and possibly the world's) altitude record for a *static* test" [emphasis added].

In 1955, a Corporal missile narrowly missed an observation post about 15 miles from the launch pad. Several minutes before every launch, people at field sites throughout the area were told to vacate their buildings so they could watch out for an errant rocket. The crew at the distant observation post did just that. As they watched, the missile began to wobble at a height of about 2,000 feet and its engines were shut down. Ron Chandler, one of the crew, described in a 2008 article what happened next: "Any missile launch always looked like it was almost over your location because of the height, so we just kept watching it come our way; however, in just a few seconds we realized that it was really coming down on top of [our location]. . . . Our quonset-type building was located on a concrete or asphalt pad that ran out about 75 feet beyond the building. I was looking over my shoulder about the time the Corporal hit the sand and exploded and sent up a dust cloud. . . . People were knocked off their feet either by the concussion or each other. . . . I hate to think what would have happened if the missile had landed another 20 feet closer and hit the concrete pad. Fortunately, the soft sand had absorbed most of the impact."

Given WSMR's nature as a testing ground, mishaps must be expected. However, safety has always been a priority, and frightening episodes like the ones just described have led to improvements in precautions and procedures. Considering the dangerous nature of rocket tests, the facility's safety record is impressive. Eckles wrote in 2006, "There have been no injuries in the general public during over 44,000 rocket and missile tests at White Sands."

Resources

Eckles, Jim. *Historic Aerospace Site: White Sands Missile Range*. American Institute of Aeronautics and Astronautics, 2004.

Levy, David H. *Clyde Tombaugh: Discoverer of Planet Pluto*. Tucson: The University of Arizona Press, 1991.

Meeter, George F. *The Holloman Story: Eyewitness Accounts of Space Age Research*. Albuquerque: The University of New Mexico Press, 1967.

Neufeld, Michael J. *Von Braun: Dreamer of Space, Engineer of War*. New York: Alfred A. Knopf, 2007.

Swenson, Loyd, James Grimwood, Charles Alexander. *This New Ocean: A History of Project Mercury*. NASA Special Publication-4201. Accessed at http://www.hq.nasa.gov/office/pao/History/SP-4201/toc.htm.

Chapter Three: Animals Aloft

The first manmade vehicle to successfully carry live passengers above the surface of the Earth was a balloon. In France, the Montgolfier brothers, who pioneered hot-air ballooning, decided it would be prudent to send live animals up into the sky before risking human life. If the animals survived unharmed, they would be reassured that humans would also be able to survive such a revolutionary journey. They sent a chicken, a duck, and a ram up 1,650 feet in September 1783. The animals survived their 2-mile, open-gondola ride without harm, except for a broken wing the chicken sustained when the ram accidentally stepped on it. All three animals were donated to Marie Antoinette's Royal Zoo, where they lived out their normal life spans.

During the following month, emboldened by the animals' test flight, several people ascended in a balloon while it was tethered to the ground. Finally, in November 1783, two men climbed into the Montgolfiers' newest balloon, stoked the fire to heat the air, and lifted off in the world's first manned free-flight voyage. The pilot, Jean-Francois Pilatre de Rozier, was a physicist and curator of the Museum of Natural History. The passenger, Marquis de Arlandes, was the leader of the French armed forces, and he flew at the order of King Louis XVI. A crowd of 500,000 people, including Benjamin Franklin, witnessed the launch. The 25-minute flight ended successfully with a landing 5 miles away, on the other side of the River Seine. Humans could now enjoy flying, and there was no more need for animal tests.

Airplanes had not yet been invented, but rockets seemed to offer the potential for faster flight than a balloon. In 1806, Claude Ruggieri, an Italian pyrotechnics expert who was living in Paris, used solid-fuel fireworks propellants to send small animals as high as 600 feet and parachute them safely back to Earth. When he announced a more powerful attempt to send a full-grown ram aloft, a young boy volunteered to substitute for the animal and become the first human blasted skyward. Ruggieri was willing, but the French police were not. They refused to allow the dangerous experiment. Ruggieri's vision of manned rocket flight would not be fulfilled for a century and a half, when liquid-fuel rockets were well developed.

On Earth, a person experiences the normal force of gravity (1 g). Acceleration or deceleration can multiply that force. At 2 g's, a 150-pound person effectively weighs 300 pounds.

New Questions

Throughout the 1800s, as technology improved, several intrepid balloonists probed the heavens. They learned that with increasing altitude, oxygen became scarce, air pressure decreased, and temperatures dropped. They also discovered that, surprisingly, the changes did not take place in a continuous way. Unmanned flights carrying instruments revealed an abrupt change in atmospheric conditions at an altitude of 6–8 miles. Human flights went as high as 23,000 feet with no special equipment, but with excruciating—and sometimes fatal—results. Gradually, the aeronauts learned how to protect themselves and bring along adequate oxygen supplies.

Gravity presented another formidable challenge for very high flight. As the distance from the Earth increases, the force of gravity decreases. Scientists and physicians wondered what a weightless state would do the human body. Would people become disoriented and unable to function? Would essential body organs fail to operate when deprived of their natural orientation and physical forces? Furthermore, both the acceleration during rocket launch and the deceleration caused by passing through increasingly thick atmospheric resistance on re-entry would significantly increase the gravitational forces on the body. Could a person withstand suddenly being many times heavier than normal?

A new, frightening discovery came in 1912, when an instrumented balloon revealed the existence of what came to be called cosmic rays. Spewed out by the sun and stars, cosmic radiation actually consists of electromagnetic waves and subatomic particles—such as bare nuclei of heavy elements like carbon and iron—traveling at nearly the speed of light. They defy protective barriers. Physicists struggled to understand these powerful forces and their potential danger. In its February 1932 issue, *Time* magazine reported that Dr. Robert Millikan (who coined the term *cosmic ray*) had found that the radiation penetrated 50 feet of lead or 200 feet of water. Research showed that the rays increased as altitude increased. The dreams of Robert Goddard, Wernher von Braun, and others faced a new challenge: would space travelers be severely injured or killed by these unstoppable particles? To some extent, weightlessness and the increased gravity forces of acceleration and deceleration could be simulated on Earth. The effects of cosmic radiation, however, could only be studied at high altitudes, and direct exposure to living organisms was the only way to evaluate the effects.

Balloons and Biology

With the end of World War II, the US military's attention began to shift from combat missions to the potential of space exploration. Balloon experiments offered

46

two important advantages for probing questions about the physical challenges of high-altitude flight. First, rockets were still under development and were not reliable enough for effective biological experimentation. More importantly, rocket flights could expose test subjects only to short periods of high-altitude conditions. Balloons, however—as old-fashioned as they might seem—presented the opportunity to attain very high altitudes and remain there for extended periods.

The Air Force, formerly the Army Air Corps, became a separate branch of the US military after the war. While the Army conducted its Operation Paperclip at the White Sands Proving Ground, the Air Force assumed a supporting role at Holloman Air Development Center on the site of the former Alamogordo Army Air Field, adjacent to WSPG. One of its missions was to investigate the biological challenges of high-altitude flight, particularly cosmic radiation. In 1947, medical researchers at Holloman launched a helium balloon that carried fruit flies to an altitude of 106 miles. Fruit flies reproduce quickly, allowing scientists to examine multiple generations in a short period of time and see whether genetic mutations appear. These high-flying flies were recovered after their capsule parachuted safely back to Earth; neither they nor generations of their descendants showed any ill effects.

Dr. David Simons came to Holloman in 1948 as project officer for animal studies. Under his direction, a wider variety of animals experienced the balloon rides. Mice were popular because they were small and easy to examine. Black mice were useful because the impact of a cosmic ray on a hair follicle would turn a black hair white. Albino mice had sensitive eyes that might be especially vulnerable to developing cataracts from radiation. Hamsters, for some reason, seemed to be more resistant to cosmic particles than other animals. As experiments became more advanced, cats, dogs, and rhesus monkeys (all anesthetized during flights) rode balloons to altitudes as high as 18 miles, with flights lasting as long as 28 hours.

In *Men, Rockets and Space Rats*, author Lloyd Mallan described in detail one flight Simons prepared and monitored. The passengers were twelve mice (six black, two brown, and four white), four hamsters, and a thermos bottle full of anesthetized fruit flies. Over a period of several hours, Simons and his assistants began placing each animal in its passenger compartment, carefully lifting each fruit fly with tweezers and placing it in the thermos bottle. Then they arranged the animal containers in a spherical capsule, along with chemicals that absorb carbon dioxide and water vapor to keep the air healthful. They also inserted several packets of film that would record cosmic particle tracks. They attached various instruments and recorders, thermal insulation, oxygen tanks, and a radio beacon to the capsule and sealed it. On the outside, they placed a mechanism that would release a parachute at the end of the flight. At dawn, they took the capsule to the launch site and affixed it to the balloon.

Simons wanted to recover the animals as soon as they landed, but it was possible that the trackers would lose contact with the balloon. A sign attached to the gondola offered a $25 reward to anyone who found it and shipped it back, postage paid. If that happened, the animals would not be alive by the time they arrived back at Holloman, but Simons would still be able to detect some useful information.

Early in the morning after the launch, Simons heard that the balloon had disappeared. He boarded an Air Force plane that would try to reconnect with the balloon's tracking beacon. They headed in the direction of its last known location, over central Texas. Knowing the balloon had been moving northeasterly, Simons told the pilot to contact the Air Force base in Shreveport, Louisiana, and ask if there had been any reports of flying saucers within several hundred miles of the base. (Because of their great altitude, such huge balloons, made of transparent plastic, reflected the sun's rays after dark and before sunrise; the eerie effect often prompted people to report seeing an unidentified flying object or UFO.) Simons was excited to hear that almost a thousand reports had come in from northeast Texas, northern Louisiana, and southern Mississippi. The plane headed in the direction the balloon was apparently traveling and eventually picked up the beacon's signal. Ultimately, Simons was able to see the balloon, release the gondola when the balloon began losing altitude, and direct a helicopter to the location where the gondola landed. By the time the capsule was picked up and brought to Simons at an Air Force base in Tennessee, three of the animals had died of heat prostration. The other seven critters and the flies survived their 30-hour flight.

This flight was one of several dozen that carried biological payloads above New Mexico between 1950 and 1956. Two flights in July and August 1954 are particularly interesting. A java monkey named Dave was a passenger on both flights. Simons sometimes sent the same animal up for more than one trip to lengthen its total time of exposure to cosmic radiation. On his two trips, Dave spent 62 hours floating at 96,000 feet. At that altitude, he was above 99 percent of the Earth's atmosphere. He survived in good health. Dave had a companion on the flights, another small monkey, who survived the trips but died in infamy a year and a half later. "The monkey learned how to unlock his cage and one night last fall he broke free and began roaming in the closed laboratory room," reported an August 1956 article in *New Mexico Magazine*. "He jerked the thermostat loose, and heaters came on." The following morning, attendants found the monkey dead from heat prostration.

Sending animals to very high altitudes for extended periods of time provided important information about cosmic radiation and opened the door for future efforts to send man into space. In a 1987 interview, Simons said, "Over a period of four or five years of this, we'd done enough experiments that it was clear there could be some [radiation] effects detectable, but they were certainly not as bad as was first prognosticated."

Animals Riding Rockets

Balloons were the best type of vehicle for cosmic radiation experiments, but they did not fully reflect the challenges of space travel. They did not reveal how living creatures would react to the sudden and severe gravitational forces of rocket launch and re-entry, or to the weightless state that would prevail between launch and re-entry. The arrival of the German V-2s at White Sands offered an opportunity to explore rockets as passenger-carrying spaceships.

The main purposes of the V-2 program at WSPG were to understand the operations and launch procedures of the rockets, and to launch atmospheric instruments and experiments in the nose cones in place of wartime explosives. Project Blossom, which sought to perfect parachute systems that would return the payloads safely to the ground, began with the twentieth V-2 launch at White Sands. Air Force engineers were developing a detachable nose cone that had somewhat more room than the nondetachable German version. In fact, the new nose cone was a prototype for an escape module being developed for the proposed X-2 piloted rocket plane. As they prepared for the first V-2 launch using the new nose cone capsule, the engineers requested a "simulated pilot" to ride the rocket. This was the project that brought Dr. Simons to Holloman.

During the next two months, Simons and Dr. James Henry, his supervisor from the Aero Medical Laboratory at Wright-Patterson Air Force Base in Dayton, Ohio, worked on preparing for the launch. The passenger would be a 9-pound rhesus monkey that came to be known as Albert. Although neither Simons nor Henry was an engineer, they decided to build a sealed capsule to contain Albert, 2 hours' worth of oxygen, and the monkey's instrumentation within the nose cone. The size and shape constraints of the nose cone made this a very difficult task, but they completed it shortly before the scheduled launch date.

On June 11, 1948, Simons and Henry brought Albert and his gear to WSPG. They decided to anesthetize the monkey. In a report the following year, Simons explained the reasons for that decision: "If the parachute opened properly, the monkey would get what was, at that point, an unknown opening shock that was calculated to be pretty impressive. . . . If the parachute streams and it comes in either half open or tangles—well, the monkey could land hard but not fatally. And we wouldn't want the monkey to be lying there suffering until we could get there. . . . I think one other thing, too, from the monkey's point of view—the thing we put them [Albert and his successors] in was a seat that they were strapped into. And they didn't have room to wiggle at all. . . . If they were awake and conscious, this would be a cruel thing to do to the animal if it wasn't necessary."

The V-2's liftoff was successful, but Albert's journey was not. Mercifully unaware of what happened during the 37-mile-high flight, Albert died. He may have suffocated because of his cramped position; heart and respiration monitors recorded no activity after he was sealed into his capsule. In addition, the parachute failed to operate properly, and Albert would have died when his container hit the ground.

"We were disturbed about the whole thing," Henry told Lloyd Mallan. Not only had they lost poor Albert, but they had not managed to collect any data about the flight's effects on the monkey. "But we had been initiated into 'rocketeering' and now knew it was a problem of achieving foolproof reliable instrumentation."

Project Blossom continued with four more launches over the next two years. Engineers devised better capsules for the next three monkey pilots (named Albert II, III, and IV), and Simons and Henry were able to gather data that showed the animals experienced no trauma from the g-forces or weightlessness in flights that reached

> "It is interesting to note that the V-2 and Aerobee aeromedical flights aroused strong complaints from certain animal lovers in the United States and abroad, but the flights also inspired a surprising number of human volunteers to write and offer themselves as passengers in the next rocket. Such offers have come to Holloman from as far away as the Philippines. Often, although not invariably, they have been made by persons hoping to pay some debt to society by gathering scientific information at considerable risk and inconvenience to themselves. One offer, in fact, was submitted in November 1956 by a resident in the Washington State Penitentiary."
>
> From *History of Research in Space Biology and Biodynamics*

heights of more than 80 miles. The final flight in the series carried a mouse that was awake and free to move about in a clear, plastic compartment. A wire mesh floor in the cage gave the mouse a surface it could cling to. A movie camera mounted in the capsule recorded the mouse's movements during the flight. The film showed that the mouse sometimes grasped the wire mesh, and sometimes floated around, apparently content. The Blossom flights were only partially successful. They produced important information, but the engineers were never able to design a parachute system that worked for the massive V-2 rockets. All of the Blossom animals died.

Meanwhile, under a military contract, Aerojet General had developed a light, slender rocket called the Aerobee. Designed as an experiment-carrying rocket rather than an explosive-carrying missile, it proved more successful than the V-2. After one failed trial at WSPG, a September 1951 Aerobee flight ended successfully with a soft landing of the capsule, which contained an anesthetized monkey (Albert VI) and eleven mice. By the time the recovery crew arrived at the landing site, however, the desert sun had overheated the metal capsule and Albert VI died on the way to the Holloman base hospital. Nine of the mice survived.

Monkey astronaut success finally came in May 1952, when Philippine macaque monkeys named Patricia and Michael (affectionately nicknamed Pat and Mike) rode an Aerobee rocket 36 miles into the sky and survived in good condition. After being examined carefully at Holloman, Pat and Mike were retired to the National Zoo in Washington, DC. About two years later, bad-tempered Mike got annoyed with Pat and bit her arm. She died of an infection caused by the bite. Mike lived until 1967.

Six monkeys and a number of smaller animals gave their lives in these early rocket voyages into the New Mexico sky, but without the lessons learned from those sacrifices, man would not have ventured into space travel. Having learned that cosmic radiation was not devastating and that living creatures could tolerate weightlessness and the gravitational forces involved in launches and landings, men were ready to take the leap.

Preparing for Project Mercury

In 1959, the National Aeronautics and Space Administration (NASA) announced Project Mercury, which was designed to send men into orbit around the Earth and return them safely to the ground. For the sake of safety and ultimate success, the daring project would proceed in several phases. Animals would get the honor of going first,

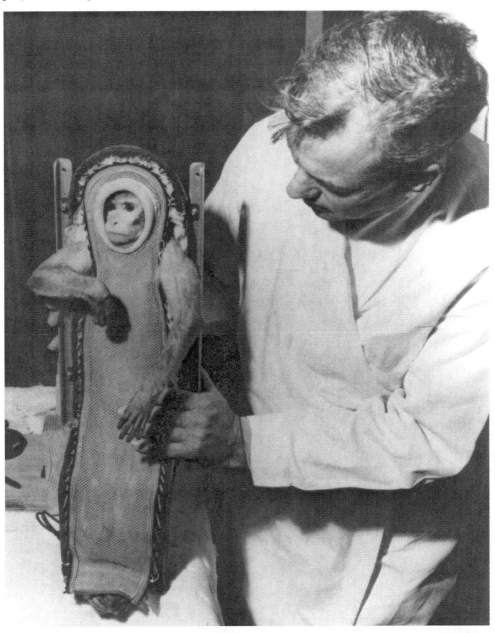

Monkeys played a role in Project Mercury, too. In December 1959 and January 1960, NASA launched Little Joe rockets from Wallops Island, Virginia, to test the Mercury capsule escape system. Each carried a rhesus monkey. Sam (an acronym for the School of Aviation Medicine) flew the first mission, and his mate, Miss Sam, flew the second. They had been trained at Holloman to pull a lever when a red light blinked on—a task they were to perform throughout their 10-minute flights. Both monkeys had successful flights and safe landings. Sam is shown here after his ocean splashdown and recovery.

making sure the equipment and procedures worked as planned. They would have to learn complex tasks, which they would be expected perform during a space flight to demonstrate whether they could function normally under flight conditions.

NASA chose chimpanzees for this project. Chimpanzees are intelligent and trainable. They can sit in human-like postures and use their hands to perform tasks. Their internal organs are arranged much like those of humans. And their reaction time (0.7 seconds) is very close to that of humans (0.5 seconds).

During the 1950s, the Aeromedical Field Laboratory at Holloman acquired sixty-five chimpanzees that had been captured in their native habitat in Africa. Because

Chang watched lights on his instrument panel and responded with appropriate lever pulls.

52

the chimps were only about a year old when they arrived, they adapted quickly to interacting with humans. Sergeant Edward Dittmer, who had been working with Simons since 1955, was named officer in charge of training the chimpanzees.

"We started out by teaching them to sit in these little metal chairs set about 4 or 5 feet apart so they couldn't play with each other," Dittmer said in a 1987 interview. "We dressed them in these little nylon web jackets which went over their chests and we could then fasten them to their chair. We'd keep them in the chairs for about 5 minutes or so and feed them apples and other fruit, and we'd progressively put them in their seats for longer periods each day. Eventually they'd just sit there all day and play quite happily."

The seats looked like small recliner chairs without the designer upholstery. Their foam-cushioned backs sloped at a comfortable angle, and elevated footrests gave the chimps an alternative to sitting cross-legged. After the animals became used to sitting in the chairs for extended periods of time, each of them was provided with a custom-fitted fiberglass couch that could be used in a space capsule. The next step of training was to get the chimps used to sitting in their couches instead of the chairs.

The final step in this first phase of chimp training was to attach a panel, with three lights and two levers, across the couch and teach each animal to respond appropriately to signals. The training used both positive and negative reinforcement. Correct responses earned the chimp a banana-flavored pellet or a sip of water from a tube mounted near its mouth. Incorrect responses resulted in a mild electric shock to the sole of its foot (the shock was unpleasant but not harmful). Dittmer and his assistants taught the chimps three basic tasks:

> Continuous avoidance. A red light on the panel stayed on all the time, and the chimp just ignored it.
>
> Discrete avoidance. A white light was above one of the levers. The chimp's task was to push that lever every 20 seconds. The white light would come on when he pushed the lever correctly.
>
> Reaction time. A blue light came on for 5 seconds about every 2 minutes. The chimp had to push the lever below it before the light went out.

It took about three weeks for a chimp to complete this first phase of training. In the second phase, which usually lasted about four weeks, the chimps learned to perform these tasks in different environments, such as an isolation chamber.

A typical work day for the chimps lasted for 6 hours, with alternating work and rest periods of half an hour each. In *The Holloman Story*, George Meeter described a session: "Four of them sat in their space chambers closely adjoining one another and their names, as I heard presently from the white-coated lieutenant in charge, were Minnie, Tiger, Elvis, and Chang. . . . old-man faces showing patiently resigned expressions as they leaned back in their small metal chairs yawning and dozing between training stints, fingering their instrument handles, pulling absently at their belts, seemingly ignoring the world."

In the third phase of training, which might last several months, a chimp learned to perform more challenging tasks. For example, in addition to the three basic tasks, he might have to watch for a green light to come on and wait 20 seconds before pulling the lever associated with it. Or he might have to solve a series of oddity problems, in which several matching shapes and one different shape were shown on a display panel. The chimp had to pull the lever beneath the non-matching shape. The chimp would get used to the oddity problems consisting of a certain group of shapes such as circles, squares, Xs, and vertical dashes. As a further challenge, a new set of shapes such as horizontal dashes and triangles might suddenly be introduced to see if he would respond appropriately to an unexpected event.

The third training phase also taught the chimps to concentrate on their tasks no matter what was going on around them. Sometimes, a chimp was seated in a mock-up of the Mercury capsule, where other instrument panels competed for his attention. Or he would be expected to perform in a simulated space chamber, where the temperature, air pressure, and humidity levels were varied to mimic space flight conditions. Trainers could play a recording of the noise of an actual rocket launch and make the chimp's capsule vibrate as it would in flight. Chimps were even challenged to perform their tasks while riding a centrifuge or a rocket sled that increased the gravitational forces on their bodies. At other times, they rode in airplanes that flew a series of vertical parabolic paths that produced weightlessness at the top of each trajectory. In many ways, the chimps' training paralleled the training of the human astronauts.

The Smart Ones

Like humans, the chimps showed various levels of ability. Some "washed out" of the program because they could not endure the long periods of training, or they could not understand or remember the tasks well enough. And some became star pupils.

The Holloman psychologists even tried to teach the chimps to count. In one experiment, they would flash a number on a screen and reward a chimp if he pressed his lever that number of times. One clever primate named George recognized numbers up to nine and pressed the lever accordingly. A variation of the task was to train chimps to press a lever exactly 50 times. A November 1961 article in the *Washington Evening Star* reported that "the chimps would pull the lever 'bangity-bangity-bang' about 45 times, then carefully pull Numbers 46, 47, 48, and 49, and finally make

---

Chimps did not get to have all the fun. Researchers at Holloman used various animals to explore how disorienting weightlessness might be. Cats were particularly interesting because of their highly developed sense of balance, which enables them to land on their feet regardless of their position at the beginning of a fall. Initial tests showed that cats did not retain this keen sense when they were weightless.

pull Number 50 with one hand cupped under the dispenser to receive the reward."

"Undoubtedly they have the power of reasoning," Jim Cook told George Meeter. Cook was in charge of the Holloman vivarium, where the animals lived. "They'll try to open doors and untie shoe laces. A chimp will use a stick like a tool to get an apple out from under something. He'll think out which rope to pull on to get food. He'll also call over his fellows and they'll 'discuss' the problem. They recognize symbolic values, colors, poker chips—they'll accept poker chips as rewards and exchange them later for some delicacy." Cook also said that "the best ones seem to work for the sheer joy of working."

One of the more interesting experiments involved teaching them to play Tic-Tac-Toe against one another. Two chimps in adjacent glass chambers each had panels that displayed the typical game board. Alternately, the panel was activated for one of the two contestants. He could press a 4-inch square on the grid to indicate his next move. "They intently eyed their squares, knuckled this or that one with a touch, leered at each other and finished each game in nonchalant triumph (with food pellet in hand) or sulking defeat," Meeter reported.

Impish Chimps

The chimps' handlers quickly learned that the animals had very distinct personalities. Some were introverts, some were extroverts, and some were downright mischievous. "Some have been known to fill their mouths with water and then spray it on an approaching attendant who had incurred their displeasure," wrote two Air Force officers in *Animal Astronauts*. "There are others who take seeming delight in hiding objects just before the technicians need them."

The chimps enjoyed playing tug-of-war in groups and squirting each other with water hoses whenever they got the chance. "One day young Rocky got the hose away from a cage handler and had a rare time squirting the man until he could be caught again," Jim Cook told Meeter. "Also they're great mimics and imitators. They play with their space jackets, take a jacket off and try it on their fellows. If one sees an attendant peeling an orange, the next time he'll peel it too."

Handlers regularly drove the chimps from their living quarters at Holloman to the Base's training facility. Chang, one of the smartest of the group, once figured out how to unlock the rear door of the van he and his classmates were being transported in. As described in *Animal Astronauts*, "He then delighted passing motorists and pedestrians by waving socially to everyone—until one of the motorists, fearing [Chang] might get hurt, signaled the driver of the van." After that episode, the handlers disabled the interior access to the door lock.

Enos, another star pupil, had a particularly earthy disposition. In *Animals in Space*, Colin Burgess and Chris Dubbs report that Enos "displayed back at Holloman a remarkable propensity for dropping his diapers and brazenly stroking his genitalia whenever reporters paid a visit to the sheet metal building where the chimpanzees were housed. . . . gleefully masturbating (which earned him the unenviable title of 'Enos the Penis')."

Enos showed another aspect of his personality in a different episode described in Guenter Wendt's autobiography, *The Unbroken Chain*. Wendt was a pad leader at NASA's launch facility in Cape Canaveral, Florida, during Project Mercury. At one point, several chimps who were candidates for an upcoming flight were sent from Holloman to the Cape for final training. During that time, a Congressman visited the facility and asked to see what he insisted on calling "the monkeys," even after Wendt explained that they were actually chimpanzees. The only animal that was not in a training session at the time was Enos. Enos was not mean, but he was not particularly sociable either. "I knew Enos would be a poor spokesman for the program so I returned to tell the Congressman that none of the chimps were available," Wendt wrote. But the politician insisted on seeing one of the "monkeys," so Wendt took him to Enos's cage. On seeing an approaching stranger, the chimp grabbed the bars and started to growl. Unabashed, the Congressman approached the cage and began teasing what he called the "little spaceman." Unamused, Enos backed away and squatted on his hands. "Enos brought out his hands from under him, gripping a steamy load of freshly laid feces," Wendt wrote. "With a snarl, he flung it straight onto the Congressman. I had seen it coming and had already backed well away." That was the Congressman's last visit to the chimps' quarters.

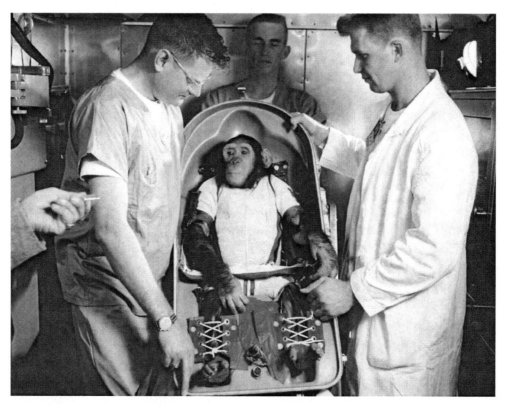

Ed Dittmer, left, had worked with Ham for about a year before strapping him in to his biopack container for the Mercury flight.

Hamming It Up

Chang, who later came to be known as Ham, was at the other end of the personality scale from Enos. "I had a great relationship with Ham," Dittmer told author Colin Burgess in a 2005 interview. "Ham was a real gentleman. He was a good animal, laughing all the time."

One of the most intelligent and trainable of the chimps, Ham accumulated more than 250 hours of training, sometimes in very stressful conditions. Safely strapped into his flight couch, Ham even rode a rocket sled that accelerated very rapidly and then stopped abruptly, simulating the forces of launch and re-entry.

Because Ham was so cooperative and well trained, he was chosen to be one of the final candidates for the first passenger-carrying Project Mercury launch. In early January 1961, Ham, another male chimp, four female chimps, and twenty human handlers and medical specialists went to Cape Canaveral to prepare for the suborbital flight. The chimps traveled and lived in two separate groups to prevent germs from spreading to all six of them if one got sick. For the same reason, a team of handlers was assigned to each group, and they had to stay away from the other group's handlers.

On January 30, the day before the launch, Dr. John Mosely (the head of the Holloman chimpanzee training program) and Dr. Henry picked Ham from one group and Minnie from the other group to be the primary and secondary candidates. The next morning, Ham was healthy, alert, and in a good mood, so he won the chance to ride the rocket. He was three years old at the time, and he weighed 37 pounds. That made him the heaviest animal sent into space to date; in the Soviet Union, several dogs had gone on suborbital and orbital flights, but each of them weighed less than 15 pounds.

Dittmer and his assistant strapped Ham into his flight couch and attached the cover, which had a window above the familiar panel of lights and levers. Then they carried this biopack container to the launch pad and secured it in the Mercury capsule atop the Redstone rocket at 7:10 a.m. "It looked like he was smiling at me," Dittmer said later.

Ham dutifully performed his tasks as he waited on the launch pad. The scheduled lift off time of 9:30 a.m. passed as engineers worked to correct a succession of minor problems. When the rocket blasted off at 11:55 a.m., Ham's wild ride began. A malfunctioning valve let too much liquid oxygen flow into the engine, so the

When Chang was chosen for the Mercury launch, his handlers changed his name to Ham. The official story is that the new name was an acronym for Holloman AeroMedical research laboratory. But Ed Dittmer told Burgess, "Our lab commander at that time was a Lieutenant Colonel Hamilton Blackshear, whose friends called him Ham, so there may have even been a dual purpose behind that particular name."

rocket accelerated more quickly than planned. As a result, the flight went 42 miles higher and 132 miles farther downrange than expected. In midflight, Ham experienced weightlessness for 6½ minutes. Then, during re-entry, he experienced g-forces as high as 14.7.

Despite the noise, vibration, and widely varying gravitational forces, Ham performed amazingly well. He had only two assigned tasks for the flight. Once every 2 minutes or so, the blue light came on, and Ham had to pull the left-hand lever within 5 seconds. He also had to pull the right-hand lever within 15 seconds of when the white light lit up. During the 17-minute flight, he made no mistakes at all in responding to the blue light. However, when the explosive separation of the Mercury capsule from the booster rocket briefly subjected Ham to a 17-g jolt, he missed two pulls of the right-hand lever. The mild electric shocks to the soles of his feet got his attention, and he did not miss any more responses to the white light.

As planned, the capsule splashed into the Atlantic Ocean at the end of the trip. However, the greater-than-expected acceleration of the spacecraft caused two problems. The capsule hit the water with so much force that it was damaged in several places, allowing water to seep in. And the rescue ship had to travel farther than expected to reach the capsule. Ham floated in the capsule for an hour before a helicopter from the ship arrived and lifted the spacecraft out of the water. Fortunately, Ham's sealed biopack container kept him dry inside the flooded capsule. The helicopter carried the capsule to the rescue ship, where the crew opened it and let Ham out.

Back at Cape Canaveral the next day, a large group of reporters and photographers wanted to see Ham. His handlers brought him out twice, but he showed his dislike for the noisy crowd and frequent bursts of flash bulbs by bearing his teeth and screaming. He flatly refused to go near a biopack couch for photographs.

Several days later, Ham returned home to Holloman, where he spent the next two years performing his usual tasks so observers could determine whether there were any delayed effects from the space flight. Ham's successful flight gave NASA officials enough confidence in the system to launch America's first astronauts into space. Three months after Ham's flight, Alan Shepard piloted a similar Mercury capsule on a 15½-minute suborbital flight. Two months after that, Gus Grissom repeated the feat.

In 1963, Ham retired from the space program and went to live at the National Zoo in Washington, DC. For the next seventeen years, he lived alone in a stark cage. Then, in 1980, he was moved to the North Carolina Zoological Park in Asheville, where he could live in a more natural environment and interact with other chimpanzees. Two and a half years later, he died quietly of heart failure. His remains are buried at the New Mexico Museum of Space History in Alamogordo, New Mexico.

Enos in Orbit

The suborbital flights of Ham and the human astronauts gave NASA—and all Americans—confidence in this country's space program. Pressure for progress remained high, though, as the Soviet Union led the way at each step of progress. Next on the agenda was a manned orbital flight. NASA, remaining cautious, wanted to demonstrate the safety of the more powerful Atlas rocket and the ability of a human-like creature to withstand the flight using American technology. That meant another chimpanzee flight.

In late October 1961, five chimp candidates from Holloman gathered at Cape Canaveral. One of them was Ham. "But, it seemed that Ham, who had already endured one ballistic flight, was not at all keen to go on a second mission," Burgess and Dubbs wrote. "He tackled his tests with little enthusiasm." Three rookies—Duane, Jim, and Rocky—remained in contention for the honor; but ultimately, Enos got the assignment. He may have been impudent, but he was also intelligent. Enos, now five and a half years old, had demonstrated his abilities in more than 1,200 hours of training, one-fourth of which was conducted with him strapped into his form-fitted couch.

A multiple-orbit flight would last several hours, so the chimp would have to pay attention during alternating periods of rest and work. Also, the duration of weightlessness would be much longer, and scientists worried about the effect that might have on his ability to concentrate and function effectively. In order to study this problem thoroughly, they devised a series of in-flight tasks that was more complex than what Ham had faced in his suborbital flight. Enos would be presented with four distinct types of problems on a rotating schedule. He would work on each task for about 12 minutes and then rest for 6 minutes before being presented with the next problem. These were his duties:

> The first problem was the same two-light set of tasks Ham had performed during his flight.
>
> In a second scenario, when a green light came on, Enos had to wait 20 seconds and then pull an associated lever. If he responded correctly, he would get a sip of water. If he pulled the lever too soon or too late, he would get nothing.
>
> For his third task, Enos had to pull a lever exactly fifty times. If he performed correctly, he would get a banana-flavored pellet. There was no penalty for incorrect performance.
>
> The fourth problem was identifying the odd symbol in a group of three that appeared on a screen. If he made a mistake, he would receive a mild electrical shock to his feet.

He was expected to not pull any levers during the rest periods.

After a few delays, the flight was set for November 29, 1961—ten months after Ham's trip. After eating breakfast, Enos was dressed and strapped into his couch. Sensors were attached to various parts of his body to monitor his physical condition. Then his biopack container was closed and placed in the Mercury capsule on the rocket. Enos waited there patiently during an hour and a half delay in the launch countdown. Then, 2 minutes before liftoff, his first problem set appeared and he began working.

This time, the rocket performed as planned and the capsule began orbiting the Earth. The flight was not without equipment malfunctions, however. During the second orbit, one of the capsule's small thrust rockets quit working. The thrust rockets were designed to nudge the bell-shaped capsule and keep it positioned correctly so the retrorockets would be pointed in the right direction when they fired to bring the capsule out of orbit and return to Earth. When one of the thrusters failed, the capsule started to tumble. Ground controllers were able to stop the tumbling long enough to end the flight safely after two orbits, although three orbits had been planned.

In the meantime, Enos was struggling with a different malfunction. During the third and fourth sessions of the odd-symbol recognition problem, the center lever stopped working. During those two sessions, he received electrical shocks to his feet for correct pulls of that lever seventy-nine times. He became frustrated, and the onboard camera showed him randomly pulling the other levers. It showed something else, too. Preflight preparations had included inserting a catheter to contain normal urination during the flight. The medical personnel had chosen a balloon catheter for this particular chimp, to keep "Enos the Penis" from removing it and distracting himself with physical stimulation. But the frustration caused by the malfunctioning lever pushed Enos over his limit of tolerance. He ripped the catheter out and soothed himself in his favorite way.

During his 50,000-mile-long flight, the well-trained Enos performed his tasks very well, except for the times the center lever failed to work. His rate of correct responses to all four of the problem types was very close to his preflight averages. This convinced NASA staff that 3 hours of continuous weightlessness would not cause disorientation or inability to function normally.

Three hours and 20 minutes after blasting off from Cape Canaveral, Enos's capsule splashed down in the Atlantic Ocean. It floated there, under aerial surveillance, for an hour and a quarter until the recovery ship arrived. A little over 3 hours later, with the capsule on board the ship, crewmen pulled the biopack container out of the capsule and released Enos. Witnesses reported that Enos jumped up and down, ran around shaking hands, and jumped into Ed Dittmer's arms. Then he ate two apples and two oranges—a welcome treat after being on a low-residue diet for days before the launch.

On February 20, 1962, three months after Enos's flight established the safety of America's orbital flight technology, astronaut John Glenn flew a similar Mercury capsule on a three-orbit mission.

Other than a mild bacterial infection caused by Enos's rough removal of the balloon catheter, the chimp suffered no physical problems from his space adventure.

Enos means *man* in both Greek and Hebrew.

After a week of examinations in Florida, he returned home to Holloman. His life there was normal until he contracted an incurable type of dysentery. He died on November 4, 1962.

Dress Rehearsal for Gemini

While Project Mercury gained momentum, NASA began planning the next phase of America's space program, Project Gemini. Its flights, which would include docking maneuvers with vehicles that were already in orbit, would last as long as two weeks. Based on their experience of testing the Mercury program equipment before manned flights, the staff of Holloman's chimpanzee training program decided to find out whether a chimp could endure a Gemini-length flight.

"A Laboratory Model for a Fourteen Day Orbital Flight with a Chimpanzee," an October 1961 report written by Holloman researchers, described their experiment. They strapped a chimp into his form-fitted couch and placed it, uncovered, in a soundproof cubicle that was 4 feet long, 2½ feet wide, and 5 feet high. Two observation windows in the cubicle walls were one-way glass so the chimp could not see out. There were elaborate mechanisms for providing fresh air and removing waste. The chimp stayed in the cubicle continuously for fourteen days, working about 9 hours each day. He had to earn all of his food and water by performing tasks like those Enos had done while in orbit. At the end of the two-week test, the chimp was in good physical condition and weighed the same as it had at the beginning. It did have a little trouble walking on weakened legs, but by the next day that problem was gone. The report gave no name for the five-year-old chimp, referring to him only as "the subject."

Project Gemini was launched in 1964, with two unmanned (or unchimpanzeed) flights to test the launch and reentry equipment. A year later, the third flight carried two human astronauts. With the era of training chimponauts over, the Air Force closed the Holloman Aeromedical Field Laboratory in 1971 and leased the buildings and remaining animals to Albany Medical College for pathology and toxicology research.

Resources

Bergwin, Clyde, and William Coleman. *Animal Astronauts: They Opened the Way to the Stars*. Englewood Cliffs, NJ: Prentice-Hall, 1963.

Burgess, Colin, and Chris Dubbs. *Animals in Space: From Research Rockets to the Space Shuttle*. New York: Springer, 2007.

History of Research in Space Biology and Biodynamics. Air Force Missile Development Center, 1958. Accessed at http://history.nasa.gov/afspbio/contents.htm.

Mallan, Lloyd. *Men, Rockets and Space Rats*. New York: Julian Messner, 1955.

Meeter, George. *The Holloman Story: Eyewitness Accounts of Space Age Research*. Albuquerque: University of New Mexico Press, 1967.

Chapter Four: Silent Skyhooks

The stratosphere, the second layer of the Earth's atmosphere, begins at an altitude of about 6 miles (32,000 feet) and extends to a height of 31 miles (164,000 feet).

Just as helium-filled balloons prepared the way for Ham's and Enos's rocket rides, they also laid the groundwork for humans to travel into space. Like silent skyhooks (imaginary hooks anchored in the sky), they had held animals in the stratosphere for days at a time, collecting crucial information that allayed fears about the dangers of cosmic radiation. In 1953, the Holloman Air Force Base's Balloon Branch began using them for even more dramatic experiments. And, again, David Simons played a major role.

Dr. John Paul Stapp, head of the Aeromedical Field Laboratory at Holloman, worked closely with Simons and knew him well. In fact, he had appointed Simons to lead the laboratory's Space Biology Branch, with its animal excursions in balloons and rockets. In August 1955, Stapp asked Simons if he thought it was time to build a bigger capsule and try a manned balloon flight into the stratosphere. Specifically, Stapp envisioned hanging the manned capsule from a skyhook that would hold it above 100,000 feet for 24 hours. At that altitude, the environment would be functionally equivalent to outer space.

Simons thought for a few moments about the ambitious proposal. He knew from the animal experiments that cosmic radiation was not a significant danger. He figured that a human-sized passenger would need a life support system with a capacity two-and-a-half times as large as they had designed for the animal flights, and he thought that could be accomplished. He also knew that recent advances in balloon technology would make it possible to carry heavier payloads. Finally, he told Stapp that he thought the project would be possible.

But Stapp had one more question. "Dave," he said, "would *you* be interested in going?"

The question surprised Simons, who considered himself more of a scientist than an adventurer. He thought a bit longer and then replied, "Well, basically, with what I know, I surely have no compunctions." He realized this would add a new dimension to his experiments. "It could very well give us a lot of information that we can't get from animals," he said. "You can make observations that the animals can't tell you about."

Project Manhigh

Stapp and Simons began planning the new program, which they called Project Man High (eventually, the last two words became one). After several unmanned and animal flights to test the new equipment and procedures, the project would culminate with manned flights. Cosmic radiation would be monitored during the flights, but the main purposes were to develop a sealed capsule that could support human life for a day or more and to demonstrate man's ability to function in the isolation of a space capsule.

The extensive development of equipment as well as the expense of the project made it too important to hinge on the availability of only one pilot. Joe Kittinger, a skilled airplane test pilot at Holloman Air Force Base, volunteered to serve as a backup to Simons. He often flew for the Aeromedical Field Laboratory, performing the vertical parabolic flights that exposed men and animals to weightlessness.

Recognizing the unique challenges the Manhigh flights would present, Stapp decided on several requirements Simons and Kittinger would have to meet. A basic physical checkup ensured that each man was in good health. Each had to pass a claustrophobia test by sitting for 24 hours in a mock-up of the closed capsule, wearing the custom-tailored, skin-tight, partial-pressure suit that he would wear during a flight. The pressurized capsule would protect the pilot from the extremely low atmospheric pressure that exists at high altitudes, but the pilot would also wear the partial-pressure suit as a backup in case the capsule lost pressure. If that happened, tubes throughout the body suit and matching gloves and socks would automatically inflate, exerting pressure on the pilot's body. The suit's helmet would also be pressurized with an oxygen-rich atmosphere so the pilot could breathe.

Simons and Kittinger got used to the pressure suit's operation during sessions in a decompression chamber that simulated the temperature and pressure of the atmo-

"Dr. Stapp knew that when we went into space, we were going to have a small capsule, a small spacecraft. We had not determined the life support systems, we had not determined the communications, the selection process, the training process. . . . Dr. Stapp came in with a program called Manhigh. This program was to investigate those things."—Joe Kittinger, Project Manhigh pilot

sphere at an altitude of 100,000 feet. At that height, the balloon would be above 99 percent of the Earth's atmosphere.

Stapp also required Simons and Kittinger to complete training for a balloon pilot's license so they could safely fly the vehicle. Simons, who had never used a parachute, also had to make at least one jump from an airplane. That way, he would know what to do if something went wrong with the balloon flight and he had to bail out.

After nearly two years of planning, training, and equipment development, the first manned flight of Project Manhigh was ready to go. Although the project was headquartered in New Mexico, Stapp, Simons, Kittinger, and other support personnel traveled to Minnesota for the launch. This flight, like many of Simons' previous animal-carrying balloons, would take off from a northern state to expose the passenger to higher concentrations of cosmic radiation in order to evaluate its effects. (The Earth's magnetic field deflects most of the charged particles that approach the Earth near the equator; with increasing distance from the equator, fewer particles are deflected.) In this case, the Minnesota launch site was also logistically convenient since it was near the Winzen Research factory, where the new capsule and the balloon were made.

Stapp decided that Kittinger, the experienced test pilot, should make the first flight and Simons would go next. This strategy placed the first flight's main emphasis on the performance of the capsule, although the pilot would also make observations and provide physical data such as blood pressure, pulse, and body temperature. Kittinger's test flight would use a smaller balloon (only 2 million cubic feet, compared with the 3 million cubic foot version that Simons would fly), stop nearly a mile below the final goal of 100,000 feet (18.9 miles), and stay there for 12 hours rather than the 24 hours planned for Simons. After the capsule and procedures were proven, Simons could focus more on scientific activities during the full-scale, second flight. "The Manhigh system as I saw it was primarily a laboratory, one in which I could conduct experiments that would be impossible in any other laboratory," Simons wrote in his 1960 book *Man High*. "I was far more interested in the frontier than the covered wagon that would take me there." Each man's interests were well matched to his assigned flight.

Manhigh I

At 12:30 a.m. on June 2, 1957, Kittinger stepped into the Manhigh capsule and began breathing its artificial atmosphere. The capsule's air supply contained more oxygen and less nitrogen than normal air, along with a good measure of helium to reduce the fire danger of an oxygen-rich mixture. Kittinger had to breathe this mixture for several hours before the launch to purge his body of nitrogen. If a person experiences a quick decrease in atmospheric pressure, the nitrogen normally dissolved in his body is released, forming bubbles in the blood and body tissues. This results in an excruciatingly painful and possibly fatal condition known as the "bends." Although the capsule would be pressurized during the flight, the pressure would be only half of that at ground level—an environment that could cause the bends.

Kittinger could breathe comfortably, but as he talked with the outside crew, they noticed that his voice became high pitched, as if he had been inhaling the helium from a birthday balloon. On subsequent flights, the pilots learned to compensate for this comical effect by speaking in unusually low voices.

By 6:30 that morning, the balloon was filled with the right amount of helium to carry the capsule aloft, and the flight began. It soon became apparent that Stapp's decision to use a test pilot for the shakedown flight was a good idea. The first problem Kittinger noticed, just after he left the ground, was that his voice transmission system did not work. He could hear Stapp and Simons talking to him, but they could not hear his replies. Using a backup system, he had to send messages to the ground by tapping out words with the dots and dashes of Morse Code. It was an inconvenience

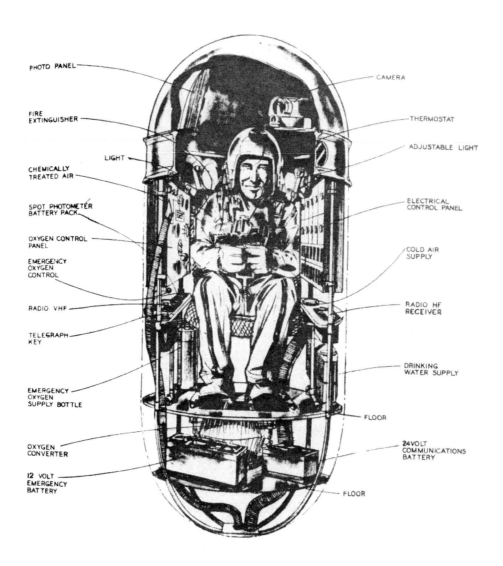

The 8-foot-high, 3-foot-diameter Manhigh capsule was packed with instruments and supplies, leaving little room for the pilot to move.

> *Ballast* is heavy items that can be thrown overboard during the flight to lighten the balloon, increasing its altitude or decreasing its rate of descent.

that limited his ability to transmit instrument readings and comments.

The next challenge was passing safely through the tropopause, a turbulent boundary layer between the near-Earth troposphere and the higher stratosphere. Manhigh I encountered this boundary at an altitude of 45,000 feet. The balloon was made of the same material as the plastic wrap people used in their kitchens—clear film only two mils (two thousandths of an inch) thick. Kittinger knew this material would become brittle as the temperature fell to 100 degrees below zero. The wind suddenly blew at 100 miles an hour, whipping the balloon around wildly. But the fragile envelope survived the violent gyrations. As it passed into the stratosphere, the temperature began rising again, the wind calmed, and the threat subsided.

When the balloon reached its ultimate height of 96,000 feet, Kittinger hovered 24,000 feet higher than any man had ever reached before. "I was really the first man in space," he said in a 2009 speech. "The sky was black. It was an eerie sensation."

But Kittinger had little time to enjoy the view. He had recently noticed a serious problem: only half of the capsule's original oxygen supply remained. Instead of releasing oxygen into the capsule, an improperly installed valve was venting it to the outside. Kittinger manipulated controls, redirecting the oxygen through the tubes in his pressure suit. This stopped the leaking, but the remaining supply was barely enough to last through the descent back to Earth. His hopes of floating aloft for 12 hours vanished. He began operating valves on the balloon to release helium and begin his descent.

"I'd been valving, but the sun was coming up and heating up the gas, so I would valve but I couldn't come down," Kittinger said. "I didn't have very much ballast, which meant that if I came down too fast, I did not have a means to stop the descent. So I was being very cautious about how I valved."

Kittinger was focusing his attention on operating the balloon, sending only brief Morse Code messages to the ground crew. Not realizing that he was already trying to bring the balloon down, Stapp and Simons sternly ordered him to start his descent. Keeping his sense of humor active in these dire circumstances, Kittinger tapped out his reply: "Come and get me." But Stapp and Simons, who were worried that the extremely high altitude was making the pilot hallucinate, didn't get the joke until later.

Stepping up his valving efforts, Kittinger finally got the balloon to start its descent after spending 2 hours at an altitude of 18 miles. His balloon pilot training paid off as he guided the vessel down through various temperature ranges that sometimes pushed it back upward. He landed successfully at one o'clock in the afternoon, just as his oxygen supply ran out.

> "In the days of Manhigh, we *were* the space program, and we were the astronauts. NASA hadn't even been formed yet."—Duke Gildenberg

Manhigh II

For two months following the Manhigh I flight, engineers and technicians worked on the capsule to prepare it for the next trip. They redesigned the oxygen system, upgraded the communications system, and installed new equipment to monitor the pilot's physical condition automatically. After exhaustive testing of the capsule and all of its components, the capsule was ready. Simons also got ready, following a low-residue diet that would enable him to endure the restriction of sitting in the capsule for more than 24 hours.

Duke Gildenberg, the Holloman team's expert meteorologist, watched the weather patterns diligently to pinpoint favorable launch conditions. This balloon was more fragile than the one used for the first Manhigh flight. The polyethylene film was only 1.5 mils thick, and the balloon was 50 percent larger. During inflation, a slight breeze would be enough to push it around, perhaps snagging and puncturing it. As a precaution, the launch site was located at the bottom of a 450-foot-deep open-pit mine whose walls would shield the balloon until it was released. Inflated with the right amount of helium for the flight, the balloon looked almost empty, with a relatively small bulge of gas at the top. Including the gondola (the Manhigh capsule) hanging below, the balloon system stretched 350 feet high. When it reached the very low pressure of the stratosphere, the balloon would swell into a 200-foot-diameter sphere.

Simons was sealed in the capsule at eleven o'clock the night of August 18, 1957, at the Winzen plant in Minneapolis. After the interior atmosphere was stabilized, the capsule was loaded onto a truck for the 140-mile drive to the mine. At the launch site, the crew attached the capsule to the balloon that was being inflated. Finally, the balloon pulled the capsule off the ground and headed skyward shortly before 9:30 in the morning. It ascended uneventfully, reaching an altitude of 101,500 feet after 2 hours and 20 minutes.

Simons noticed the absolute silence of space. "It's like no Earthly quiet," he said. "On Earth there are always traffic sounds and dogs barking or the wind just whistling. But in space there's nothing but quiet."

During nearly two years of preparing for his flight, Simons had planned two dozen experiments to do in the capsule. For example, he brought along a 5-inch telescope so he could observe the moon and Venus without the distortion of the Earth's atmosphere. He planned to study cloud formations from high above. And he would monitor his own physical and psychological performance in unique circumstances and events. He had scheduled free periods of time, too, but he was so fascinated by his surroundings that he chose to take photographs and gaze through the telescope during these relaxation breaks.

The problems Kittinger experienced in the test flight had been solved, but Simons soon found new challenges to deal with. The communications system became erratic and needed frequent adjustment. A radio beacon that allowed the ground crew to track the balloon failed. The temperature inside the capsule varied widely. Daytime highs around 80 degrees made him perspire in his snug pressure suit. As the nighttime temperature fell to 32 degrees, he struggled in the confined space to put on an extra layer of clothing. He became so busy with his tasks and enthralled with the views of Earth and sky that he forgot to eat anything. During their radio conversations, Stapp recognized the symptoms of low energy and loss of concentration, and he reminded Simons several times to eat something.

After sunset, Simons enjoyed the stellar view. "I knew there would be no twinkling . . . because the turbulence of the atmosphere was no longer present," he said later. "But the blue stars were bright, clean blue. The red stars were bright, clean red. It was as if, for the first time, I saw the sky in color."

Soon, however, his attention was distracted by a dangerous development. Without the sun's warmth, the helium in the balloon cooled more than expected. As it cooled, the gas in the balloon shrank, and Simons began losing altitude. He rapidly approached storm clouds that had gathered below him. If he got too close to them and lightening struck the 200-foot-long antenna that dangled below the capsule, the jolt would burst the balloon. Periodically throughout the night, Simons dropped ballast to lighten the load and allow the balloon to rise a little higher. He dozed occasionally, but could not sleep soundly because of his cramped position and his concern about falling into the storm. About six o'clock, the sun rose and the balloon slowly began to warm and climb higher.

The daylight and a hearty breakfast (cold ham, eggs, sliced peaches, and a nut roll) cheered Simons. "I felt as if I no longer belonged to the Earth on this morning," he wrote in *Man High*. "My identity was with the darkness above. As I ate, the sky around me and above me grew darker. I knew that I was returning to the altitudes I had visited the day before. It was right." At the time, he did not realize that he was experiencing the *breakaway phenomenon*, a euphoric attachment to space that could make a person not want to return to Earth. It was what Simons had feared Kittinger was experiencing in the previous flight when he said, "Come and get me."

As the hours passed, the balloon rose higher. By 10:30, it floated at its maximum altitude of nearly 102,000 feet. Simons continued making observations and relaying instrument readings to the ground crew. Stapp began noticing that Simons was making mistakes and not concentrating well. He asked the pilot to check the carbon dioxide concentration in the capsule, and they discovered that it was at a dangerous level. The cold temperatures during the night had decreased the efficiency of the

"It is a forlorn waste, but across the vast, unfertile expanse there exists a sense of space and grandeur matched only by the deep blue of the sky above, which seems to beckon man into the greater cosmic wilderness."—David Simons

chemical filter that was designed to absorb the carbon dioxide Simons exhaled. Stapp told Simons to close his helmet and breathe the oxygen supplied through his pressure suit. This quickly restored his mental alertness. It also kept additional carbon dioxide from being released into the capsule. The filter system, which had warmed up and was now operating at full strength, was able to remove the excess buildup from the capsule, and Simons was able to reopen his helmet.

With only enough battery power left to last through the descent, Simons began venting helium from the balloon at noon. Five and a half hours later, he landed safely in an alfalfa field in South Dakota. A helicopter that had been tracking his balloon promptly arrived to take him back to Minneapolis.

Mission Accomplished

Although each of the Manhigh flights had encountered some difficulties, the program largely accomplished its goals. The pilots satisfied a set of selection criteria and fulfilled their prescribed training requirements. Kittinger's test flight proved that it was possible to build a space capsule that could support human life for extended periods of time. Refinement of the design would improve its reliability and length of operation, but his successful ascent and landing showed that it could work.

Simons' flight proved that a man could endure the physical confinement and psychological isolation of the capsule during long flights. His flight lasted just over 32 hours, but he was actually sealed in the capsule for 44 hours. He felt some discomfort sitting that long in the cramped space; but with some guidance from the ground crew, he was able to function effectively even under the stress of life-threatening problems.

Project Manhigh had produced useful results, but it had overspent its budget and could not afford more flights. The Holloman team headed back to New Mexico to work on something else.

Manhigh Revived

Stapp, Simons, and Kittinger had extraterrestrial flight in mind while they worked on Project Manhigh. Their superiors in the Air Force, however, still considered space travel a comic-book fantasy not worthy of substantial research funding. Their thinking changed abruptly only six weeks after the Manhigh II flight. On October 4, 1957, the Soviet Union put the world's first artificial satellite in orbit. *Sputnik* might as well have meant "fund space research." The US government wanted to put a man in space before their Cold War rivals did. Project Manhigh had made important progress in developing and operating a space capsule. The Holloman Aeromedical Laboratory received money to revive the project to continue that work for another flight.

Kittinger and Simons had used the same capsule, but Winzen Research built a new model for Manhigh III. It looked a lot like the previous version, but it used a new chemical life-support system to remove both carbon dioxide and water vapor from the air. It also used a different strategy for temperature control.

The difficulties Kittinger and Simons faced during their flights emphasized the importance of having skilled pilots who could remain calm during physical and emotional stress. "An essential quality necessary to an astronaut would be stamina, not in a purely physical sense but in a psychophysiological sense," Simons wrote in *Man High*. "A combination of deep physical reserves plus the all-important emotional determination to use those reserves."

An important objective of Manhigh III was to develop procedures for evaluating candidates—not only for the pilot of this balloon flight, but for future astronauts who would ride rockets into space. The requirements Kittinger and Simons had fulfilled formed the basis of the new criteria. New candidates would also have to undergo four days of physical testing at the Lovelace medical clinic in Albuquerque, complete a full day of psychological tests and interviews with a psychiatrist, and endure several physiological stress tests such as spending an hour in a "hot box" with a temperature of 155 degrees and humidity of 85 percent.

Six candidates were considered for Manhigh III. Four of them were dropped for various reasons. The healthier and higher performing of the remaining two became the primary pilot, with the other being his backup. The primary pilot was Clifton McClure, a 26-year-old Air Force pilot with a master's degree in ceramics engineering and an interest in astronomy. McClure was so fascinated with the Aeromedical Field Laboratory's space biology work that he had arranged to be assigned to Holloman, where he was willing to do any job just to be near the action.

"It was supposed to be a detriment to your career to get assigned there [the desolate Holloman]," McClure told reporters before his Manhigh flight. "But I didn't care what I had to do. I would've carried garbage out in the morning just to be there."

Volunteering for the Manhigh tests brought him right into the action. "I started getting myself evaluated along with the rest of these people. And I beat 'em every time," he said. "That was my objective."

With the equipment and the pilot finally ready, the team went to Minnesota in late September 1958. But fall, with its unfavorable weather, was closing in. A scheduled launch was canceled because of poor conditions. Time pressure became intense. Earlier that year, Congress had established the National Aeronautics and Space Administration (NASA), a civilian organization charged with conducting the nation's manned space program. The Air Force was allowed to continue with the third Manhigh flight only because the equipment had been paid for and the training was complete. Waiting for better weather in the spring or summer was not an option.

During a preflight press conference, a reporter in Minnesota asked McClure why he volunteered for Manhigh. "The way I look at it is this: I was brought up to believe in the country I live in. I feel you have to put more into it than you take out of it if it's going to continue as it is." Then he added, "I'm afraid I'm not afraid. Anything of this sort has its serious side, but I think it's also going to be quite a bit of fun."

The only realistic solution was to return to New Mexico to launch the flight. The weather would not be a serious problem. In fact, the wind patterns would probably keep the balloon over the vast White Sands Missile Range throughout its flight, simplifying the capsule recovery process. The unparalleled WSMR tracking facilities would add a greater level of security to the mission. Cosmic radiation exposure would be reduced, but measuring its effects was not a major part of this third flight, anyway. Within a week, the equipment and personnel were in Alamogordo preparing for the launch.

Early on the morning of October 7, the ground crew carefully unfolded the fragile balloon and laid it out on a runway at Holloman Air Force Base. McClure had already been sealed in the capsule for several hours when it was attached to the balloon. When the balloon was inflated almost enough to take off, an unexpected gust of wind tossed the balloon around. It bounced into the ground and burst. One chance remained for the final Manhigh flight. Winzen had made two balloons in case one failed. The launch was rescheduled for the following morning.

After getting 7 hours of sleep during the day, McClure ate a low-residue dinner of steak and potatoes. Then he put on his pressure suit and re-entered the capsule shortly after midnight. About 3 hours later, he was asked to report the oxygen pressure in the capsule. As he turned to read the gauge, he brushed against the pilot's emergency parachute, which hung in front of him. The parachute popped open, filling his lap with a pile of loose fabric.

For several minutes, McClure tried to figure out what to do. If the capsule was opened so the parachute could be repacked, it would delay the launch for several hours. By then, it would be too windy, and the launch would have to be postponed. The weather forecast was not favorable for the next several days. Most importantly, McClure did not know whether the ground crew had already started to lay the balloon out on the runway. Because the balloon could be damaged so easily, it could not be repacked. And it was the last balloon they had. McClure decided to try to repack the parachute himself, inside the capsule, without telling anyone about the problem.

Several months earlier, he had happened to see someone packing a parachute. As usual, he was curious about anything he had never seen before. He watched intently and asked questions about the process. Now, he concentrated to remember what he had learned. It was hard to move in the cramped capsule, and there was very little room to manipulate the wads of fabric. After working diligently for 2 hours, he finally managed to get the properly folded parachute packed back into its pouch. He was exhausted from the effort.

"I finished just after leaving the Aeromedical Field Laboratory building on the truck for the launching site," McClure wrote in a post-flight report. "I was soaking wet with sweat and began to be very cold as I lowered my activity level and the capsule cooled in the early morning air. . . . I lay over against the chute, now stowed in its normal place, and attempted to sleep."

Outside the capsule, launch preparations proceeded smoothly. The balloon inflated without any wind problems. Shortly before seven o'clock, Manhigh III rose into the air. Between performing his flight-related duties, McClure watched out the six porthole windows as he floated into the sky. A pair of mirrors mounted outside one of the portholes gave him views straight up and straight down.

"The view through the down mirror was amazingly clear during the ascent, and the entire flight, for that matter," McClure wrote in his post-flight report. "Through this mirror could be seen the light white-brown to reddish colors of the desert, the small specks that were the mesquite and other desert vegetation, the sharp jagged streaks that marked the paths of the dry stream beds, and other details of similar dimension and contrast."

By ten o'clock, the balloon floated at its maximum altitude of 99,600 feet. From that vantage point, McClure could see into outer space. "I see the most fantastic thing, the sky that you described," he radioed to Simons. "It's blacker than black, but it's saturated with blue like you said. . . . I'm looking at it, but it seems more like I'm *feeling* it. It's literally indescribable!"

Around noon, McClure suddenly realized he was hungry. "Immediately with this thought came the consciousness of fairly strong hunger feelings and also *extreme* thirst," he wrote. "I made note of the fact that the engrossment of capsule preparation, launch, climb, and being at such an altitude had caused me to forget completely such normal procedures as eating and drinking." Except for a small can of juice he drank a couple of hours before takeoff, he had not eaten or drunk anything for 15 hours. He decided to take time out and eat a meal. This was the first flight to carry foods that were designed to be eaten in a weightless environment. He selected a toothpaste-type tube containing three and a half ounces of creamed chicken, a six-ounce can of juice, and two Oreo cookies. He also began to use the capsule's dispensing system to drink water.

The food relieved his hunger, but McClure continued to notice that he tired too easily when doing simple tasks. During radio conversations, Simons noticed that McClure's speech was sluggish. Monitors showed the pilot's pulse rate higher than it should be. McClure reported feeling too warm. He was perspiring heavily, but it did not help him cool off because his pressure suit trapped the moisture against his body and kept it from evaporating. At 1:30, the command crew asked him for temperature readings. He reported that the air temperature in the capsule was 96 degrees and his body temperature was 102.3 degrees.

If he continued to float at that altitude, the sun would continue to heat the capsule for many hours. McClure also noticed that the air blowing out of the re-generation system was very hot; the chemical process that removed water vapor and

During Manhigh II, Stapp had to remind Simons to eat. Because of that and McClure's similar experience, NASA would include food and liquid consumption in the activity schedules for manned space missions.

carbon dioxide from the air produced heat. Unless the cabin temperature dropped significantly, McClure would not survive. His supervisors ordered him to start valving helium from the balloon and begin his descent.

To ensure a safe landing, the descent could not be too fast. As the balloon slowly drifted lower, McClure's body temperature continued to rise. He kept his activity level as low as possible to conserve energy. By three o'clock, the balloon had come down to 87,000 feet, but McClure's temperature had risen to 104.1 degrees. The crew monitoring his condition ordered him to buckle his safety harness; if he lost consciousness from heat exhaustion, he would not be able to control the descent and the landing impact would be hard. But being buckled up restricted his movement and increased his physical discomfort.

McClure did manage to remain conscious through the entire descent. He made a well-controlled soft landing on the flat desert at seven o'clock in the evening. Just before he opened the capsule to climb out, he recorded his body temperature as 108.5 degrees. The recovery helicopter had been tracking his descent and landed moments later. The crew rushed over to help him out of the capsule, but he was determined to climb out himself.

"I had to insist that I be allowed to jump down, and upon doing so found that this jolt made my knees shake and gave a light pain in my stomach," McClure wrote. "It was a complete relief to remove my helmet and only then did I get a chance to feel how hot I really was. The cool night air provided a tremendous comparison to the air I had felt for the last few hours."

The crew rushed McClure to the base hospital at Holloman. By the next morning, he had completely recovered from his ordeal.

McClure's performance under extreme physical and emotional stress was truly remarkable. When Joe Kittinger heard what had happened, he said, "He's the only one who could've survived. He's just a tough, tough guy."

Manhigh officials emphasized the importance of the rigorous pilot selection process. Eli Beeding, the project physiologist for Manhigh III, wrote in his post-flight report, "Although our extensive pilot selection program, which included reactions to stress, was not perfect, it did serve its function, in that our pilot, in spite of the physical state discussed, brought his cumbersome flying machine in after dark for a perfect landing."

Project Excelsior

Six months before the third Manhigh flight, Stapp had moved to Ohio to head the Aero Medical Laboratory at Wright-Patterson Air Force Base. He invited Kittinger, who was no longer working on Manhigh, to go with him and work on high-altitude escape systems that could be used either for aircraft or for spacecraft during launch or re-entry. Parachutes worked effectively at altitudes below about 20,000 feet. But opening a parachute at a much higher altitude would leave the jumper descending slowly through freezing temperatures and low atmospheric pressures, with insufficient oxygen to breathe safely.

Beginning in 1954, researchers in the Air Force's Project High Dive tried to find a way to let a jumper fall quickly through the hostile high altitudes and then open a parachute in a lower, more survivable atmosphere. Using various modifications of parachutes and procedures for opening them, anthropomorphic (physically human-like) dummies wearing parachutes were dropped from airplanes and from balloons at altitudes as high as 100,000 feet. High Dive was headquartered at Wright Field in Ohio but conducted its balloon tests at Holloman to take advantage of the Balloon Branch's experience, the good weather, White Sands' open expanse, and the extensive tracking and retrieval capabilities. Tests showed that if a dummy was allowed to free fall to an altitude safe for parachute use, it would fall in a horizontal position and begin to spin like a propeller. In some experiments, this "flat spin" reached 200 revolutions per minute—a speed of whirling that would certainly cause unconsciousness and probably cause death for a person. The High Dive researchers tried various ways of preventing flat spin but were unable to find a solution. The project was canceled in 1957.

When Kittinger arrived at Wright Field in the spring of 1958, he reviewed the experiments and results of High Dive. Then he went to Stapp and asked for permission to revive the project. Stapp agreed and authorized Project Excelsior, led by Kittinger.

One of the members of the Excelsior team was Francis Beaupre, a civilian employee at Wright Field who had previously served as a parachute rigger in the Navy. For years, Beaupre had experimented with modifications to the commonly used parachutes. Now he began working on a sophisticated parachute system that would stabilize a body during a quick fall through the upper atmosphere and then open a standard parachute for a slower descent at lower altitudes. The revolutionary system he invented is known as the Beaupre Multi-Stage Parachute.

Beaupre's system operated automatically. When a person jumped, a timer would begin counting. After 16 seconds, the jumper would not yet have started into a flat spin, but he would have a high enough air speed for a small parachute to fill. At this point, the timer would release an 18-inch-diameter pilot chute, propelling it into the air with a spring-loaded device. As the pilot chute filled with air, it would pull a 5-foot-diameter "drogue" chute out of the pack. This small chute would not slow the jumper's descent, but it would make him aerodynamically stable, holding him in a vertical position and preventing the troublesome spinning. As the jumper reached an altitude of 18,000 feet, the thicker atmosphere would have begun to reduce his speed, and a pressure-sensitive mechanism would release the 28-foot-diameter main chute without causing too severe a jolt. This large parachute would then bring him safely to the ground.

Excelsior is from Latin meaning ever higher.

Wearing his protective gear and equipment for the Excelsior flights, Kittinger weighed 320 pounds—twice his normal weight.

In tests using anthropomorphic dummies, Beaupre's new parachute system performed well. But Kittinger knew that was not conclusive enough. The dummies simulated the size, shape, and weight distribution of a man; but they could not move and maneuver like a live person. Kittinger was just the guy, with his parachuting experience and motivation, to test the new system on himself.

Kittinger returned to Holloman for his live jumps. Tests began with him and three other men first testing the parachute systems by bailing out of an airplane at 30,000 feet. Everything worked perfectly, so Kittinger began preparing for a series of jumps from high-altitude balloons. To make it easier to jump, he would ride in an open gondola rather than the closed, pressurized capsules used in Project Manhigh.

Excelsior I

The first balloon test took place the morning of November 16, 1959. To make sure that Kittinger would be over the target area at White Sands when he jumped, the balloon was launched about 60 miles northwest of Holloman, near Truth or Consequences, New Mexico.

When Kittinger arrived at the launch site, he began to dress in the many layers of clothing that would protect him from the low temperatures and pressures during the balloon ascent and the parachute descent. A nylon vest held body-monitoring sensors against his body. Next, he put on two layers of long, thermal underwear and a pair of gloves and socks. Then he squeezed into his partial-pressure suit, including its inflatable socks and gloves. Next came a thick layer of long underwear and electrically heated socks and gloves. With more help from his assistants, he struggled into a winter-weight flight suit, wool socks, and rubber boots designed for Arctic use. Then came the pressure suit helmet, the bulky multi-stage parachute backpack, and a chest pack containing an emergency backup parachute. Finally, an instrument box was strapped to his rear, where it would serve as a seat in the gondola. The 60-pound instrument box carried such items as his oxygen supply, a tape recorder, a movie camera, and batteries.

Kittinger climbed into the gondola and settled his instrument box-seat onto a Styrofoam base that contained eighteen plastic bottles. As the water in the bottles froze during the ascent, they would release heat for Kittinger's comfort. The liftoff and ascent went smoothly, although the sun overheated the balloon and lifted it to 76,000 feet—nearly 3 miles higher than planned.

When he was ready to jump, Kittinger tried to stand up, but he couldn't. The frozen water bottles had expanded so much that his instrument box was wedged in tightly. He pulled and pulled, and finally jerked it loose. Then, standing in the gondola's doorway, he pulled on the cord to start the parachute system's timer. The cord failed to pull free, so he jerked it two more times, finally freeing it. He stepped out of the gondola.

At first, he had the startling sensation that he was not falling. There was no whistle of wind or rippling of his garments. But he rolled onto his back and saw the balloon appearing to zoom away from him, although he was actually falling quickly away from it. Then he realized the parachute system was not working as planned. Unknown to him, his first tug on the cord had actually started the timer. As a result, the pilot chute sprung out of his backpack only 2 seconds after he jumped, rather than the intended 16 seconds. He was not falling fast enough for it to fill and pull out the drogue chute. Instead, it flapped around and its lines wrapped around Kittinger's neck. Without the stabilizing effect of the chute, he started to spin, quickly reaching a rate of 120 revolutions per minute.

"At first I thought I might retard the free spin that began to envelop me, but despite my efforts I whirled faster and faster," he said later. "Soon I knew there was nothing I could do. I thought this was the end. I began to pray, and then I lost consciousness."

Fortunately, his emergency parachute opened automatically at 11,000 feet and dropped him safely to the ground on the White Sands Missile Range. He regained consciousness as the recovery helicopter landed.

> "He'll outwork you and out-think you. He just stays with it and stays with it."—Alan Shepard, the first American to ride a rocket into space, describing Kittinger

After Kittinger jumped from the gondola, it was released from the balloon and returned to the ground by parachute. Another recovery crew retrieved it and the instruments it carried. The expendable balloon drifted off; eventually it would flutter harmlessly to the ground.

Excelsior II

Despite the near disaster, Kittinger firmly believed that Beaupre's parachute system would have worked except for the timer glitch. And he intended to prove it with his next jump. After three weeks of equipment preparation, he took off again from Truth or Consequences.

As the balloon reached its planned height of 74,700 feet, Kittinger stood up. Modifications to the Styrofoam seat base kept his equipment box from sticking. When he pulled on the timer cord, it released properly. He jumped.

This time, the multi-stage parachute worked perfectly. Kittinger remained vertical, with no spinning. He landed without needing his emergency chute. Now it was time to prepare for the ultimate test: a jump from higher than 100,000 feet.

Excelsior III

The final Excelsior flight was scheduled for August 16, 1960. It would not turn out to be as flawless as the previous test.

To accommodate the summer wind patterns and land in the target area, the launch site was moved to an abandoned airstrip near Tularosa, New Mexico. Around midnight, twenty vehicles paraded through Alamogordo, carrying the crew and the equipment to the launch site.

At two o'clock, Kittinger arrived at the landing strip by helicopter. At 3:30, he started pulling on the multiple layers of clothing in a van. Getting bundled up as before, he had to be surrounded by cool air to keep him from sweating before the flight. A generator-powered air conditioner kept the van's temperature at 50 degrees. Just as Kittinger was starting to dress, the flow of cool air stopped. Outside the van, a member of the ground crew heard the generator sputter. He knew that the balloon was already being inflated and the flight would be canceled if the van's temperature rose. He quickly looked inside the generator and saw that a ground wire had come loose. It was located in the back of the compartment, but reaching past the moving parts of the hot engine, he was able to grab the loose wire and hold it against the contact post. He knelt there, holding the wire in place at arm's length, for 2 hours to keep the air conditioner running.

Meanwhile, Duke Gildenberg was in another van, watching the weather. He could not tell if a nearby storm would come their way. He ordered a half-hour delay so he could watch its path. At the originally scheduled launch time of five o'clock, Kittinger climbed into the gondola to prepare for launch. Just as the hold expired at 5:30, Gildenberg reluctantly decided to cancel the flight. As luck would have it, the automatic release on the balloon tripped a minute too soon, and the balloon rose against his orders.

Time and money were running out for Project Excelsior. This was make it or

"Actually, after all the time going up I was rather glad to jump," Kittinger said at a press conference 3 hours after he stepped out of the gondola.

79

break it time. Kittinger was committed to completing the mission, and with the weather conditions marginal, not critical, Gildenberg did not suggest aborting the flight.

As the balloon climbed past 40,000 feet, Kittinger's pressure suit inflated automatically. Almost. "At 43,000 feet . . . my right hand does not feel normal," he wrote in a *National Geographic* article later that year. "I examine the pressure glove; its air bladder is not inflating. The prospect of exposing my hand to the near-vacuum of peak altitude causes me some concern. From my previous experiences, I know that the hand will swell, lose most of its circulation, and cause extreme pain. I also know, however, that I can still operate the gondola, since all the controls can be manipulated by the flick of a switch or a nudge of the hand. I am acutely aware of all the faith, sweat, and work that are riding with me on this mission. I decide to continue the ascent, without notifying ground control of my difficulty."

Now he was approaching the turbulent tropopause. "There the balloon's polyethylene fabric—only two-thousandths of an inch thick and of the same filmy material used to contain . . . dry-cleaned clothes—will become almost brittle from the cold," he wrote. "About half of balloon failures occur at the tropopause. The temperature drops steadily until it reached -94°F at 50,000 feet, then starts to rise. I have safely passed the tropopause barrier."

An hour and a half after leaving the ground, Kittinger floated at 102,800 feet. On one side, the sun heated his body, making him perspire. On the side away from the sun, he could see steam-like vapor flowing out into the cold emptiness. Looking down, he was disappointed to see that a thick layer of clouds drifted below him, blocking his view of the Earth. Although the ground crew could not see him, they tracked him by radar. Gildenberg calculated his descent path and told him to sit in the gondola for 11 minutes, drifting into position over the 11-mile-square landing target.

"I want to describe my impressions of this high, alien world," he wrote in *National Geographic*. "Striving for the right words, I send a message to ground control: 'There is a hostile sky above me. Man will never conquer space. He may live in it, but he will never conquer it. The sky above is void and very black and very hostile.'"

When it was time to jump, he was ready. He started a camera mounted on the gondola to film his fall. He radioed to the ground the information that his right hand was not pressurized. Then he jettisoned the long antenna that hung below the gondola, terminating communication with his crew. Now completely alone, he said a quiet prayer: "Lord, take care of me now." Then he stepped over the edge.

"When I stepped over I fell on my right side for about 8 seconds, then on my back looking up at the balloon and that black sky for about 10 seconds," Kittinger later recalled. "After that the stabilizing chute opened—I looked up and saw it. There was absolutely perfect control during descent." During the 18 seconds before the drogue chute opened, Kittinger fell 1¼ miles.

The drogue chute did its job as intended; it stabilized his fall but did not decrease his speed. During a total of 4 minutes and 38 seconds of free fall, he reached a speed of 614 miles per hour. Some sources erroneously report that he broke the sound barrier; but the speed of sound varies with temperature, and in the cold atmosphere he was falling through, he reached Mach 0.9 (90 percent of the speed of sound). Nevertheless, he set a record for the fastest speed a human has traveled without a vehicle. That was not the point of any of his jumps, though. He insists that setting records was never the objective; these were scientific experiments designed to ensure the safety of pilots and astronauts.

As he continued to fall, he saw the thick layer of clouds getting closer. Not having parachuted through clouds before, he had to remind himself that it was merely vapor, not a solid barrier. He entered the clouds at 21,000 feet. At 17,500 feet, still in the clouds, his main parachute opened. He uttered another quiet prayer, "Thank you, God, thank you." As he fell out of the clouds at 15,000 feet, he was delighted to see two helicopters circling nearby.

Preparing to land, he struggled to detach the equipment box. He could not use his right hand, which was swollen to twice its normal size. Unable to free one of the connections, he would have to land with the box dangling at his side. He reached the ground, 27 miles west of Tularosa at 7:26 a.m., 13 minutes and 45 seconds after he stepped out of the gondola. He hit the ground hard, and the box bruised his leg severely. "But I am on the ground, apparently in one piece," he wrote. "I am surrounded by sand, salt grass, and sage, but no Garden of Eden could look more beautiful."

The recovery helicopters landed, and three men—including Francis Beaupre—rushed to Kittinger's side. As they helped remove his helmet, he smiled and said, "I'm very glad to be back with you all." Within 3 hours, the swelling in his right hand went down and there was no lasting damage.

By ten o'clock that morning, Kittinger was back at Holloman, freshened up and dressed in his Air Force dress uniform, in time for a press conference. He said, "I was the lucky one—I made the jump. But without all these ground people we couldn't have done it at all. We tested everything we could first, and then this human test was necessary. It showed us, among other things, that man's only limit is his imagination."

The Excelsior program made significant contributions to NASA's new space program. The first astronauts were selected for the Mercury program in April 1959, less than a year after the last Excelsior flight. One of them, Gordon Cooper, summarized the importance of Kittinger's program in 1988, when he told a reporter: "It was absolutely vital. We had to know if we could build the right kind of equipment to sustain life. . . . We didn't have any idea about the body's stability at high altitudes or what kind of dynamics the human body would go through or how to build a drogue chute that would stabilize it. . . . Everybody who came along conjectured a different theory. . . . You had to send a guy up to do it."

NASA applied many of the results of the Manhigh and Excelsior programs to their space flight procedures and equipment. One thing they did not use was the Beaupre Multi-Stage Parachute system as part of an emergency escape system. In a

2003 interview for *Avweb.com*, Kittinger explained his belief that such a system could have saved astronauts' lives: "The Space Shuttle is the first experimental aircraft ever built without an escape system for the difficult portion of the flight [takeoff and re-entry]. In 1986 when the Challenger booster blew up, they peaked at 100,000 feet, and with an escape system like the one we had developed forty years ago, they could have jumped or ejected from that spacecraft and had the same ride down that I had. But they didn't have the equipment. I believe the crew was still alive inside the capsule when it hit the water, and they might have survived if they had been able to eject."

In the aftermath of the Challenger disaster, a Presidential commission studied the accident. The commission's final report included the statement that "Ejection seats (which afforded only limited protection during first stage) were provided for the two-man crews of the Orbital Flight Test program (the first four Shuttle flights). Other options for 'operational' flights carrying crews of five or more astronauts were considered, but were not implemented because of limited utility, technical complexity and excessive cost in dollars, weight or schedule delays."

Astronaut Robert "Hoot" Gibson had a somewhat different interpretation. In a 2005 interview, he said, "We were playing this game back then at NASA that said 'Oh this thing is gonna be so safe, that we don't even need parachutes.' After we lost the Challenger, we thought 'well we had that all wrong.'"

NASA reported that in 1988, "a new crew escape system was added that allows the Space Shuttle crew to bail out if the orbiter has to make an emergency return descent and a safe runway cannot be reached." The system relied on the Shuttle gliding down to an altitude of 22,000 feet, at which point a 10-foot pole would be extended out the side of the vehicle. In turn, the astronauts would slide down the pole, which would place them clear of potential contact with the Shuttle, and they would parachute to the ground.

In 2010, as the Space Shuttles were nearing retirement, NASA conducted tests at White Sands on an escape system that could be used on the next-generation, Orion spacecraft. The astronauts would remain in an enclosed capsule that could be ejected from the rocket and carried to the ground by a cluster of parachutes. In the full test, the capsule successfully detached from the launch vehicle a mile above the ground and parachuted to a soft landing. "The performance was absolutely astounding," Reed Johnson, NASA's program manager, told an *Albuquerque Journal* reporter.

Resources

"Col. Joe Kittinger," viewed at http://www.avweb.com/news/profiles/183671-1.html.

Highest Step in the World, The. Foolish Earthling Productions, DVD, 2003.

Kennedy, Gregory. *Touching Space: The Story of Project Manhigh.* Atglen, PA: Schiffer Military History, 2007.

Kittinger, Joe, and Craig Ryan. *Come Up and Get Me: An Autobiography of Colonel Joe Kittinger.* Albuquerque: University of New Mexico Press, 2010.

Kittinger, Joe. "The Long, Lonely Leap," *National Geographic*, December 1960

Kittinger, Joe. "The Sky Is My Office," presented at the National Museum of the Air Force, April 22, 2009. Accessed at http://www.nationalmuseum.af.mil/shared/media/document/AFD-090811-059.mp3.

Manhigh III: USAF Manned Balloon Flight into the Stratosphere. Holloman Air Force Base, 1961. Accessed at http://www.dtic.mil/cgi-bin/GetTRDoc?AD=AD259635&Location=U2&doc=GetTRDoc.pdf.

Ryan, Craig. *The Pre-Astronauts: Manned Ballooning on the Threshold of Space.* Annapolis: Naval Institute Press, 1995.

Chapter Five: The Human Factor

In 1950, when the V-2 rocket scientists and engineers left White Sands, the United States military was beginning to develop new generations of rockets and missiles. These costly devices were often damaged on impact after a test firing, making it not only difficult to analyze their performance but also expensive to replace them.

In an attempt to provide nondestructive tests of rockets, the engineers at Holloman Air Force Base built a test track. It consisted of a 3,550-foot-long pair of railroad rails and a wheeled carriage on which a rocket could be mounted horizontally. Firing the rocket's engine sent the carriage-mounted missile zooming down the track. A braking system near the end of the run brought the rocket to a safe stop. A 1959 *New Mexico Magazine* article quoted Colonel Donald Vlcek, chief of Holloman's Track Division, on the usefulness of the facility: "With this rocket sled track, you can go to the far end and find out why a missile failed. You don't have to dig it out of the ground and wonder."

Man on the Fast Track

When John Paul Stapp came to Holloman in 1953, he found an additional use for the test track. During a previous assignment at Muroc Air Force Base (later renamed Edwards AFB) in California, Stapp had begun a series of experiments to evaluate the effects of high g-forces on the human body. Scientists at the time were concerned that a person would not be able to withstand the severe gravitational forces produced by acceleration during a rocket launch or deceleration during re-entry into the Earth's atmosphere.

"Experiments with a rocket-propelled sled on a horizontal track on the ground may pioneer the way for an early fulfillment of the vision of human flight in interplanetary space, and certainly reduces such proposals from the status of fantastic dreams to feasible realities," Stapp told the author of a 1955 book, *Men, Rockets and Space Rats*.

Captain Edward Murphy was in charge of measuring the forces Stapp experienced in various parts of his body during the sled runs. Before one run, an assistant wired the sensors backward, so they measured zero during the jarring run. Annoyed, Murphy remarked, "If there's any way he can do it wrong, he will." As the story spread around the base, the remark became, "If anything can go wrong, it will." At a press conference a few weeks later, Stapp credited the project's safety record to their consciousness of "Murphy's Law."

While at Muroc, Stapp developed a rocket-powered sled for the base's 2,000-foot-long test track. The sled, dubbed the Gee Whiz, was equipped with a chair. A dummy strapped into the chair would ride the sled, sustaining a build-up of g's during acceleration and then suddenly undergoing high g's in the opposite direction during a very rapid deceleration. Besides measuring the forces' effects on the dummy, Stapp had to develop a system of restraining straps to hold the dummy securely without exerting pressures that would injure a human subject.

After three dozen tests with the dummies, Stapp was satisfied that the sled and the restraint system were reliable. In December 1947, he rode the sled himself. Cautiously, he instructed the crew to activate only one of the rocket engines on the sled. He reached a speed of 90 miles an hour before abruptly slamming to a stop. Finding the run not too bad, he climbed back onto the sled the next day and ordered three rockets fired. This run blasted him to a speed of almost 200 miles an hour before jerking him to a standstill.

Having demonstrated the survivability of the test runs, Stapp sought volunteers to participate in the experiments. He chose eleven airmen who satisfied him that they were not "thrill seekers." He limited tests to midweek so the subjects would not be distracted by thoughts of weekend activities. Stapp, himself, rode the sled a total of sixteen times, enduring as much as 35 g's (forces against his body 35 times as strong as the Earth's gravity). This shattered the then-theoretical human limit of 18 g's. He did not sustain any lasting injuries, although he did suffer bruises, scrapes, broken ribs, and concussions. Fillings were even dislodged from several of his teeth.

With his assignment to Holloman, Stapp had access to a test track that was almost twice as long as the one at Muroc. He seized this opportunity to extend his research and ordered construction of a much more powerful rocket sled, which he called Sonic Wind Number 1. After a series of shakedown runs using dummies and chimpanzees, Stapp climbed aboard the sled on March 19, 1954. As he stepped into the sled's chair, he told a reporter, "I assure you, I'm not looking forward to this."

One *stapp* is the force exerted by 1 g acting on the human body for 1 second. An astronaut experiencing 3 g's for 15 seconds has endured 45 stapps.

Stapp wrote *Stapp's Law*: The universal aptitude for ineptitude makes any human accomplishment an incredible miracle.

A 1955 *Time Magazine* article reported Stapp's response when a friend asked what he thought about as he waited for the sled to lurch forward: "First, I look around at the mountains and at the bright skies and I don't think about anything. Then I say to myself, 'Paul, it's been a good life.'"

With only six of the sled's twelve 4,500-pound-thrust rockets firing, Stapp reached a speed of 421 miles an hour—a new land speed record. During the deceleration, he momentarily endured 22 g's, and he experienced 15 g's for more than half a second—twice as long as he could have on Muroc's shorter track—without lasting effects.

"I feel fine," Stapp said after the run. "This sled is going to be a wonderful test instrument. I'm ready to do it again this afternoon."

In fact, his next ride on Sonic Wind Number 1 did not take place until August 20.   After modifications to the sled, this run subjected Stapp not only to severe g-forces, but also to an abrupt wind blast such as an ejecting pilot might encounter. During the run, a section of the solid windshield popped open. This time, he reached a speed of 502 miles an hour and suffered no substantial effects from the wind blast or the deceleration.

Stapp began preparing for his next, even more challenging sled run, which would use nine of the sled's rocket engines. "As the g's got higher," he said in the 2003 film

Stapp never asked a volunteer to experience something he, himself, had not already endured.

Highest Step in the World, "I started practicing dressing and undressing with the lights out, in case anything really happened to my eyes. I'd at least be in practice taking care of myself."

On December 10, 1954, Stapp followed his normal morning routine, rising at four o'clock and having coffee and an orange for breakfast. He read some medical journals and then drove to his office at Holloman, where he joked around with assistants. As Dr. David Simons started a physical exam, Stapp said, "I can tell you in advance that my blood pressure is high; my pulse is also fast due to an excessively rapid heartbeat caused by the adrenaline of anxiety being pumped into the heart at an abnormal rate." Measurements confirmed his prediction.

After dressing in a T-shirt and flight suit coveralls, he climbed onto Sonic Wind Number 1 and pulled on a helmet. Crew members strapped his helmet and his chest to the sled's chair, and tied down his arms and legs in several places to keep them from flailing uncontrollably during the run. They inserted a tooth guard into his mouth, closed his helmet faceplate, and left for protective cover.

"You feel depressed and tense waiting for the ride to begin, especially when everybody draws back to the blockhouse to fire the rockets and you wait all alone," Stapp said in a 1956 *New Mexico Magazine* article. "Everything's been worked out very well. You know that. But something could go wrong."

Suddenly, the rockets ignited, spewing 35-foot-long flames behind the sled. After 5 seconds, he was traveling 632 miles an hour—90 percent of the speed of sound.

Joe Kittinger watched in amazement from a unique vantage point. He had been assigned to fly a photographer in a jet airplane at low altitude to record the sled's run. He had practiced for days to time his arrival at the track's starting point just as the countdown ended. On the crucial day, he got it just right, then watched as the sled gained speed. "I will never forget, as long as I live," Kittinger wrote in *The Long Lonely Leap*, "the incredulous awe I felt at that moment as Colonel Stapp accelerated like a bullet away from my own speeding airplane."

After the 5-second acceleration, Stapp coasted for three-tenths of a second while experiencing 10 g's of backward force. Then the sled hit the brakes. It took a little more than 1 second to come to a standstill. During that time, Stapp endured at least 25 g's of forward force, with a momentary peak of 46 g's.

As the sled abruptly began its severe deceleration, Stapp said his vision became like a "shimmering salmon" and the pain in his eyes felt "somewhat like the extraction of a molar without an anaesthetic."

A recovery crew and an ambulance were waiting at the end of the track. They hurried to Stapp, removed his helmet, and cut his restraining straps. George Nichols, the project manager representing the sled's manufacturer, rushed to Stapp's side. "When I got up to the sled, I saw his eyes," he told a reporter in 2003. "Just horrible. His eyes . . . were completely filled with blood."

Stapp couldn't see a thing. He tried to push his eyelids up, but found that his eyes were already open. He said, "This time, I get the white cane and the seeing eye dog."

"They put me on a stretcher and in a minute or two I saw some blue specks," Stapp said in the 1955 *Time* article. "In about eight minutes or so after the stopping of the sled, the blue specks became constant and pretty soon they became blue sky and clouds. I saw one of the surgeons wiggling his fingers at me and I was able to count them." Most of the capillaries in his eyes had burst, but his retinas had not detached. "I began to feel relief and a wondrous elation," he said. "The lights were turned back on again. I could see."

Stapp later recalled having experienced some mental confusion. "In the ambulance riding to the hospital for a checkup, I began telling the story of the ride," he said in a 1956 *New Mexico Magazine* article. "The doctors were listening carefully, as they all do. I suddenly realized I was telling the story backward and shut up."

After forty minutes of examinations at the Holloman Base hospital, Stapp enjoyed two full servings of lunch. He stayed in the base hospital for five days of observation. He had two black eyes, caused by his eyeballs slamming against the eyelids during the sled's deceleration. He had some bruises and abrasions from the restraining straps. His voice was hoarse, and his sinuses were blocked for two days. During the high-speed run, tiny grains of sand had shot through his clothing and left small blood blisters on his skin.

The December 1954 Sonic Wind ride made Stapp even more famous than his previous record-setting runs. *Life* and *Time* magazines featured his rocket sled ride. He was surprised to be the star of an episode of the television program *This Is Your Life*. But fame was not his motivation.

"He did that [run] because we were looking at ejections at high speeds," Kittinger said in a 2009 speech. "We were looking at deceleration from space flight. All of these things were information that our Air Force needed and our space program was going to need."

"I want to emphasize this ride was not done as a stunt or to set a speed record," Stapp said in the 1956 *New Mexico Magazine* article. "It showed what man can stand in forces more than twenty-five times gravity. It showed what he could stand in the rate, magnitude, and duration of such forces. He's awfully tough, a lot tougher than the machines he rides in."

Undaunted, Stapp began planning for his next sled run, when he intended to travel 1,000 miles an hour. His superiors intervened, however. Deciding that his knowledge and his leadership abilities were too important to risk anymore, they ordered him not to ride the rocket sled again. "Colonel Stapp has developed a tremendous experience and know-how," said Brigadier General Marvin Demler, who was with the Air Force's Air Research and Development Command in Washington, DC. "We don't believe he should stretch his luck any farther." Demler added, "[Stapp] didn't like it one bit."

HHSTT

HHSTT is not the sound of the rocket sled speeding by, it is the acronym for the Holloman High Speed Test Track. Since the track was a twinkle in the engineers' eyes back in 1948, the HHSTT has been rebuilt and lengthened several times. Originally 3,550 feet long, it was extended to 5,071 feet in 1956, to 35,095 feet in 1957, to 35,588 feet in 1966, and to 50,788 feet in 1974. A lighter weight sled, Sonic Wind Number 2, was used for windblast experiments in 1955–1956. Other carriages and sleds have been used to carry a variety of experimental payloads such as entire rockets, nose cones, astronaut capsule components, and several kinds of animals. Ham and Enos rode the track to prepare for their space flights.

It is the longest, flattest, and fastest test track in the world. In January 2008, a classified Navy rocket rode three test track sleds that fired in sequence. The rocket traveled 3.6 miles down the track in 6 seconds, setting a new land speed record of 6,589 miles per hour.

With powerful thrusts propelling sleds at extremely high speeds, the track must be very flat and smooth. Any bumps or vibrations can literally knock the sled off the tracks, hurtling it disastrously through the air. It takes remarkable construction techniques to accomplish the smoothness. "The tolerance could vary no more than one-tenth the thickness of an ordinary pin sideways or up and down," Track Division Chief Vlcek said. "We needed very exact alignment so the sled could move at very high speed with very little vibration."

The 1957 reconstruction process illustrates the precision of building the track flat and smooth enough. At that time, 39-foot-long rails were welded together to create a 10,000-foot-long section. With this section in place on the track's concrete foundation, one end was anchored in place while powerful jacks pulled on the other end, stretching it so tight that it would not deform in 120-degree heat. After anchoring that end securely, the next 10,000-foot section was welded to it, and stretched. Ultimately, just over 35,000 feet of rail was stretched a total of more than 23 feet.

The HHSTT's braking system consists of a 1-foot-deep water trough between the track's twin rails. A scoop on the bottom of the sled drags through the water to slow down. A series of boards perpendicular to the rails act as dams to keep the water level at a desired depth. The scoop breaks the boards as it passes, increasing the sled's deceleration. Water pools can be placed at any location along the track to accommodate test runs of a desired length, and the water depth and number of boards determine the rate of deceleration.

The water brake system works very well, but it also presents a problem. Birds and other small animals see it as an appealing oasis in the desert. With the experimental cargo traveling faster than the speed of sound, they do not have time to get out of the way before the sled hits. Not only is this fatal for the animals, but it can damage the vehicle. In one case, a 212-pound sled traveling 3,000 miles an hour hit a small bird, tearing a 10-inch hole through one-quarter-inch-thick, armor-plated steel on the vehicle.

The track's operators have tried various methods for shooing away the critters just before a test run. They tried scaring them with the sound of machine gun fire or a swooping hawk, but the birds returned too quickly. Two solutions that have been pretty effective are blasting a detonation cord beside the track just before the run, and sending a small monorail sled slightly ahead of the actual test vehicle to startle the birds off the rail and out of the water just in time.

Pushing Over Daisies

Stapp's enforced retirement from the HHSTT did not end his adventures. He was still able to ride another device he had designed for Holloman's testing facilities. Built in 1955, the Daisy Track was a scaled-down version of the high-speed test track. Its rails were 5 feet apart, rather than the larger track's 7-foot separation. And it was only 120 feet long (the length was doubled in 1962). The sled was propelled by compressed air, rather than rocket engines. In fact, the track was named after the Daisy air rifle (BB gun) that was popular at the time.

Citing its "low operating costs compared to sleds propelled by rockets on large tracks," Stapp told Air Force publicist George Meeter that "the cost of desert air compressed by an electric motor stayed at less than five cents per run." The entire cost of a run was in the range of $100–150, compared with several thousand dollars for a run on the HHSTT.

Being smaller and less expensive did not mean the Daisy Track was wimpy. Its sled, running on oiled rails, could reach a speed of 100 miles an hour after traveling only 40 feet. When a piston on the front of the sled impacted a water-filled cylinder at the end of the track, it produced deceleration rates comparable to those achieved on the long track, although their durations were not as long.

"Peak g-forces ranged from 10 to 36 g's for our human volunteers and up to 170 g's on animals and dummies," Stapp told Meeter. "I might add that our human subjects seemed to find the sample blood-drawing more objectionable than the sled impacts."

One test run, however, left the human subject with a different opinion. On May 16, 1958, Eli Beeding climbed on the seat of the Daisy Track sled, facing to the rear. It was a routine test. Through previous runs of sled decelerations up to 35 g's, the researchers had learned that the chest of the passenger sitting above the sled consistently experienced one and one-half times the g-force as the sled itself sustained during the peak of deceleration. This run was designed for a maximum sled deceleration of 40 g's. Beeding, then, would experience 60 g's, but only for a very short duration of four one-hundredths of a second.

At the end of the short ride, the sled jerked to a stop, taking only a tenth of a second to decelerate from 35 miles per hour. Beeding felt an unexpectedly crushing sensation. "The first thing I felt was the spinal pain, as if a baseball bat had been applied to the bottom of the back," he told Meeter. "Then there was the numbness and dizziness. I couldn't feel my back at all. I kept trying to talk. I wanted to tell them what was happening to me, but then I sensed my vision beginning to go. I could hear the talking around me, about the blood pressure, for instance, just no pressure at all, zero. But there was too much dizziness and then I knew I was actually going to pass out. I think that's the last thing I did tell them."

The doctor and other attendants supervising the test rushed to Beeding's side.

Concerned that his back might be broken, they were very cautious while moving him into a position that lowered his head. This helped him regain consciousness. He tried to smile at them as they moved him into the ambulance. In ten minutes he recovered completely, except for a sore back.

When the observers checked the gauge that had been attached to Beeding's chest during the test, they found he had experienced 83 g's, pushing his apparent body weight to 11,620 pounds. Repeating the test with dummies verified that at a 40-g deceleration, the effect on a body jumped to *twice* the effect on the sled—a relationship no one could have predicted. Fortunately, this result was discovered during a rear-facing sled run. If Beeding had been facing forward, he probably would not have survived.

Daisy Track runs continued until 1985. Among the devices tested on the track were space capsule seats and helmets for the Mercury, Gemini, and Apollo missions. "At the cost of a few stiff backs, kinked necks, bruised elbows, and occasional profanity, the Apollo capsule has been made safe for the three astronauts who will have perils enough and left over in the unknown hazards of the first flight to the moon," Stapp told Meeter.

No Weight

Stapp, Beeding, and many other volunteers experienced crushing forces on the HHSTT and the Daisy Track. Holloman researchers also explored the opposite end of the spectrum—the weightlessness astronauts would experience in space. The terms subgravity, microgravity, and zero gravity also refer to this condition.

The first technique used to give animals and people the experience of zero gravity was to fly an airplane in a Keplerian trajectory. The pilot flies the plane in a vertical parabolic path, much like riding a roller coaster car up a steep incline and then dropping into a steep descent. At the top of the airplane's path, the centrifugal force of its motion equals the pull of gravity, creating a weightless state. This difficult maneuver requires a highly skilled pilot, particularly to maximize the period of weightlessness.

"When I heard about the zero gravity program I volunteered, and that's how I met Dr. John Paul Stapp, who was in charge of the Aeromed Lab," Kittinger said in 2009. "I did over 500 runs in zero gravity. They usually lasted 35 or 40 seconds."

Three of Kittinger's frequent passengers on these flights were the subgravity program's project officer, David Simons; Simons' chief assistant at the Space Biology Branch, Ed Dittmer; and the subgravity program's task scientist, Grover Schock. Schock had been a strong candidate for the Manhigh III balloon flight until he was badly injured during the crash landing of a helium balloon training flight.

The Holloman flights were neither the first nor the only zero-gravity experiments being done at the time, but they were among the most productive. Some results were

> "I always volunteer. Ninety percent of everything I did in life happened because I volunteered for it."—Joe Kittinger

quite unexpected. For example, during one of the early flights with the passenger seated behind the pilot, they were unable to communicate effectively with each other because the airplane's standard microphone system did not work in subgravity conditions. Engineers were able to develop a new type of microphone system that worked well. As a result, Schock wrote in a 1957 report, "Voice communications in future space vehicles should present no problem."

With an assortment of volunteers riding in subgravity flights, researchers discovered that some people really enjoyed the experience, while others were uncomfortable or even physically ill. This suggested that a good tolerance for weightlessness should be one of the selection criteria for astronauts.

On some flights, passengers experimented with eating and drinking. They learned that swallowing required new skills. The tongue had to be used to push the liquid or chewed food to the very back of the mouth to trigger the swallowing reflex. They also discovered that liquids had to be contained in squeeze bottles to keep them from spilling around the cabin. "A meal served the ordinary way would float all over the cockpit," Schock said in a 1959 *New Mexico Magazine* article. "We must know enough about this subject so that when problems arise in space travel, we will know something about how to cope with them."

Researchers also tested the volunteers to see whether their sensory and motor skills would deteriorate during weightlessness. They successfully performed tasks such as touching their finger to the tip of their nose or marking Xs in a series of squares.

The subgravity airplane maneuvers produced valuable information, but they had an important limitation. It was physically impossible to maintain zero gravity for a minute or more. But Schock wanted to study the effects of longer periods of continuous weightlessness. He thought he could simulate subgravity at ground level by submersing a person in water. The water pressure, although barely perceptible, would be evenly distributed over the entire body surface. He expected this to feel very much like weightlessness.

Schock arranged to use the indoor swimming pool at the School for the Visually Handicapped in nearby Alamogordo. He had an experimental device built to test a subject's sense of direction under water. The device had a seat, where the volunteer would sit, completely submerged in the deep end of the swimming pool. The chair could be rotated around a pole that ran up the back of the chair. The chair could also be tilted by moving that pole. The person sitting in the chair, breathing from an air tank, was blindfolded. Very slowly, the researchers would tilt and rotate the chair. They found that the subject did not notice the change of position as he would have with visual and gravitational clues. In one experiment, for example, volunteers did not realize they were leaning until the chair was tilted to a 22-degree angle. In another experiment, the researchers tilted the chair upside down and back, several times, stopping at some intermediate angle. Then they asked the blindfolded person in the chair to point which way and how much they should tilt the chair so he would be sitting straight up. Rarely did a subject end up vertical; sometimes he would be nearly horizontal when he finished giving his tilting instructions.

Schock's experiments led to a better understanding of the importance of visual clues to a person's ability to maintain spatial orientation while weightless. His report on his subgravity experiments helped earn him a doctorate in space physiology—the first in the United States.

Astronaut Selection

By late 1958, the United States' manned space program was moving at a fast pace. The federal government created the National Aeronautics and Space Administration (NASA), a civilian agency, to run the program. Among the first tasks were figuring out exactly what the astronauts' duties would be, establishing criteria for evaluating candidates, and selecting the first group of American space travelers.

A committee that included representatives of the engineering and medical professions decided that the astronauts would have five general responsibilities. A 1965 NASA report listed those responsibilities:

1. To survive; that is, to demonstrate the ability of man to fly in space and to return safely.
2. To perform; that is, to demonstrate man's capacity to act usefully under conditions of space flight.
3. To serve as a backup for the automatic controls and instrumentation; that is, to add reliability to the system.
4. To serve as a scientific observer; that is, to go beyond what instruments and satellites can observe and report.
5. To serve as an engineering observer and, acting as a true test pilot, to improve the flight system and its components.

The Manhigh balloon project had already developed some selection criteria. Using these as a starting point, an aeromedical committee developed a set of minimum qualifications for candidates, which are listed in a 1989 NASA publication:

1. Age - less than 40
2. Height - less than 5 feet, 11 inches
3. Excellent physical condition
4. Bachelor's degree or equivalent
5. Graduate of test pilot school
6. 1,500 hours total flying time
7. Qualified jet pilot

To expedite the selection process, President Eisenhower decided to limit the pool of applicants to the men who were currently serving as military test pilots. Of the 508 test pilots in all four branches of the armed services, a records review revealed that 110 of them met the seven basic criteria. Initial interviews with two-thirds of them produced 32 acceptable volunteers, which was a large enough pool of candidates.

Randy Lovelace, a doctor in Albuquerque, New Mexico, participated in developing the list of duties and the minimum qualifications for candidates. And he would play a prominent role in the next phase—physical examinations and physiological tests—of astronaut candidate evaluations for Project Mercury, NASA's program to put a man in orbit around the Earth.

The Lovelace Connection

William Randolph Lovelace II (Randy) was six months old when his family moved to New Mexico in 1908. He was named after his uncle, a doctor who had recently moved from Missouri to New Mexico, settling in Fort Sumner (then called Sunnyside) for relief of his tuberculosis symptoms. The uncle moved to Albuquerque in 1913, and in 1922 he and a partner formed a medical practice that would become known as the Lovelace Clinic. Randy lived on his father's ranches while he was growing up but stayed with his uncle in Albuquerque while he was in high school.

After attending college in Missouri and medical school at Cornell and Harvard, Randy went to work for the Mayo Clinic in Rochester, New York. While he was there, the Air Force asked the Mayo Clinic staff to develop an oxygen mask that could be used during high-altitude flights. Working with two other doctors, Lovelace helped design the BLB (the L standing for Lovelace) mask that made high-altitude aviation possible.

During World War II, Lovelace served in the Army Air Corps as head of the Oxygen Branch at the Aero Medical Laboratory in Ohio. He tackled a new problem related to high-altitude flight. People who had to bail out were dying from lack of oxygen before their parachutes brought them to the ground. Lovelace developed a breathing system using a small oxygen bottle that could be worn during the descent. In 1943, he made the only parachute jump of his life to show that the system would work. He jumped out of a bomber at 40,200 feet with the bottle taped to his leg. Although the jolt of the opening parachute knocked him out, the oxygen system sustained him, and he regained consciousness before he landed. His jump demonstrated not only the effectiveness of his invention, but also the difficulty of parachute operation at high altitudes.

With the war over and his military service concluded, Randy Lovelace moved back to Albuquerque in 1946 and joined the staff of the Lovelace Clinic. He founded the Lovelace Foundation for Medical Education and Research. Under his leadership during the 1950s, the Foundation conducted many government-sponsored research projects related to aerospace research. When NASA came into existence in 1958, he was named immediately to head its Special Life Sciences Committee. It was in this capacity that he helped develop the astronaut duties and selection criteria.

It is not surprising that NASA chose the Lovelace Clinic to design and conduct the exhaustive physical examinations and physiological tests for the Mercury astronaut candidates. It was the preeminent aerospace medical organization in the country. The Lovelace Foundation had cosponsored three national conferences on physics and medicine in the upper atmosphere. Some of the Clinic's research projects had dealt with examining and amassing data on test pilots. The staff had proven its reliability on government assignments, not only in the quality of their work, but also in their ability to maintain secrecy when necessary.

Beginning in 1955, the Lovelace Clinic screened candidates and monitored the health of pilots flying U-2 spy planes for the Central Intelligence Agency (CIA). In this secret project, pilots flew specially designed aircraft at 70,000 feet over unfriendly countries to photograph military installations. The project leaders thought that flying at such a high altitude would make the planes invisible to radar and unreachable by enemy aircraft or missiles. The Clinic's ability to manage secret projects was relevant to the Mercury project because NASA wanted the candidate testing to be confidential until it was ready to announce the chosen astronauts.

Astronaut Exams

Following their extensive interviews in Washington, DC, the acceptable candidates would undergo two phases of testing: physical examinations in New Mexico, then psychological and physical stress tests in at Wright Field in Ohio.

In accord with their top-secret orders, the first group of six candidates arrived in Albuquerque on February 7, 1959, dressed in nondescript civilian suits. The off-the-beaten-path location and small town character also helped keep the astronaut candidates out of the national limelight during their exams. With a population of 200,000, the city was not an insignificant hamlet; but its remoteness, desert environment, and Earth-hugging architecture surprised many who came to visit or to work. Dr. Donald Kilgore, who had moved to Albuquerque in 1953 to join the staff of the Lovelace Clinic, once called it the "brownest, dustiest place in the country." He described what he and his wife saw as they passed through Tijeras Canyon and drove toward the city on Route 66: "In the far distance we could see this dusty little town. There was no green anywhere."

But the astronaut candidates were not in town to enjoy the ambiance. They would spend the next seven and a half days undergoing what has been described as "one of the toughest medical examinations in history." Appointments with various medical specialists began at seven o'clock in the morning and ended at six o'clock each afternoon, except for three days when the tests ran late into the evening. There was a relentless series of X-rays, blood tests, urine and stool samples, enemas, electro-encephalograph measurements of brain activity, lung capacity measurements, treadmill and stationary bicycle activities, and much more. In one test to evaluate stomach acids, each candidate had to swallow a length of rubber tubing. Each candidate even flew to Los Alamos National Laboratory for body radiation and potassium measurements.

A U-2 flight over Cuba in 1964 revealed the construction of missile bases that were being outfitted with nuclear warheads. This led to the Cuban Missile Crisis, a two-week showdown between the United States and the Soviet Union, that brought the Cold War to the brink of nuclear war.

Nominally, each candidate underwent thirty different tests. But the regimen was not that simple—the eye exam, alone, entailed seventeen separate procedures. Being active military test pilots, the candidates were all in excellent health, so the tests were designed not to disqualify anyone but to assess their relative capabilities to endure the stresses they might encounter during space travel. For example, the Lovelace researchers devised a test to predict a vulnerability to dizziness or motion sickness. They placed calibrated goggles over the candidate's eyes and poured a stream of cold water into his ear for 30 seconds. This caused rapid, involuntary movement of the eyes. They measured how quickly the movement began, how severe it was, and how long it lasted.

Perhaps the most dramatic and unpleasant test involved electrical measurement of muscle activity. Randy Lovelace wrote an article for the April 20, 1959, issue of *Life* magazine, in which he described the test: "To test each pilot's neuromuscular system, Dr. Amick stimulated nerves in the arm to see how fast and how well the message was transmitted to the muscles. The doctor then inserted a tiny electrode in a hand muscle and measured the electrical response of the muscles to nerve stimulation. Finally, to test fatigue when the blood supply is inadequate, he cut off circulation in one arm until the pilot felt that it was "falling asleep." Dr. Amick then stimulated the nerves with repeated shocks which made the pilot's fist clench involuntarily in a sustained and rather painful way."

Candidates who experienced this test described it in somewhat more emphatic terms. Author Tom Wolfe dramatized it in his 1979 book, *The Right Stuff:* "They brought him into a room and strapped his hand down to a table, palm up. They brought out an ugly-looking needle attached to an electrical wire. . . . they drove the needle into the big muscle at the base of his thumb. It hurt like a bastard. . . . The wire from the needle led to what looked like a doorbell. They pushed the buzzer. Conrad looked down, and his hand . . . was balling up into a fist and springing open and balling up into a fist and springing open and balling up into a fist and springing open and balling up into a fist and springing open at an absolutely furious rate, faster than he could have ever made it do on its own, and there seemed to be nothing that he, with his own mind and his own central nervous system, could do to stop his own hand or even slow it down."

At the end of their week of testing, each group of candidates had a chance to meet the next group that was arriving for their turn at the battery of tests. They greeted each other with typical test-pilot humor. "Part of the ritual was an unsolicited, straight-faced indoctrination lecture which the outgoing group gave the newcomers," Lovelace wrote. "In this lecture all hypodermic needles became square and blunt, proctoscopes took on the proportions of the Palomar telescope, and all enemas and blood samples were ocean-sized."

> "They went into every opening on the human body as far as they could go."—Astronaut John Glenn

Performing the tests on the thirty-two candidates was only part of the Lovelace Clinic staff's job. They also had to organize and analyze the large amount of information generated for each candidate. Their ability to do this quickly and thoroughly, using the new technology of mainframe computers, had been an additional factor in the Clinic's selection as the test center. The entire process of testing and data analysis was completed within two months. On April 9, 1959, NASA officials presented to the American public the seven astronauts who would participate in the Mercury project: Scott Carpenter, Gordon Cooper, John Glenn, Gus Grissom, Wally Schirra, Alan Shepard, and Deke Slayton.

Lovelace and the Ladies

All of the Mercury astronauts were men. That was an automatic result of the requirement that the candidates had to be military test pilots, a position women were not eligible to hold. In that era, many Americans considered women the "weaker sex" and thought they should be protected rather than being subjected to hazardous situations.

Randy Lovelace, however, wondered whether women might be capable of space travel. Moreover, he saw potential advantages over male astronauts. Women were typically smaller than men, so they would require less space, less food, less oxygen to breathe, and less fuel to propel them into space. Women were less susceptible to heart attacks and, therefore, might be better equipped to endure the cardiovascular stresses of a space flight. Women's internal reproductive organs could be less vulnerable to radiation encountered during a flight. Some research suggested that women could endure cramped spaces and protracted periods of isolation better than men.

Donald Flickinger, a high-ranking officer at the Air Force's Air Research and Development Command, served on NASA's Life Sciences Committee with Lovelace. Together, they planned to test female candidates at the Albuquerque clinic. In September 1959, Flickinger and Lovelace attended an Air Force Association meeting in Florida, where they happened to meet another attendee who could help launch their project. She was Geraldine (Jerrie) Cobb, a 28-year-old civilian pilot who held numerous world records in aviation. Cobb enthusiastically accepted their invitation to be the first woman to undergo the astronaut qualification tests.

In December, however, Flickinger bowed to his superiors' opposition and canceled the Air Force's sponsorship of the testing program. Still personally supportive of the project, he suggested that the Lovelace Foundation take responsibility for funding it. Lovelace agreed and named it the Woman in Space Program.

Cobb reported to the Albuquerque clinic on Valentine's Day in 1960. She underwent exactly the same battery of tests the male candidates had taken, except that a gynecological exam replaced the prostate exam. In August, Lovelace announced the results at an international space symposium in Stockholm, and a week later *Life* published a photo essay about Cobb and her tests. The article summarized the results this way: "After a series of exhaustive and exhausting medical tests, seventy-five in all, during which she complained less than the Mercury men had, Jerrie Cobb easily passed the rigid requirements laid down for astronauts-in-training."

Cobb's success generated a wave of public excitement about women in space. The government sector, however, did not echo the enthusiasm. They denied outright discrimination, but tried to downplay the notion of female astronauts on practical grounds. "Women won't go into space because NASA doesn't have spacesuits to accommodate their particular biological needs and functions," Flickinger said. He

Jerrie Cobb posed for a photo beside a Mercury space capsule, but NASA never considered her a potential astronaut.

explained that "we could not justify the expense of altering the [partial pressure suits] to fit the girls."

Undeterred, Lovelace decided to recruit more women for testing, and Cobb helped identify candidates. She reviewed the records of 782 female commercial pilots, looking for the right combination of health, age, experience, accomplishment, and determination. Ultimately, Lovelace sent written invitations to twenty-five well-qualified women. Several of them had twice as many hours of flight experience as the Mercury Seven's average. Eighteen of the invitees were willing and able to participate in the Woman in Space Program.

During the spring and summer of 1961, the women came to Albuquerque to experience the same tests Cobb had passed so successfully. While the male candidates had enjoyed the camaraderie of several of their fellow test pilots during their examinations at Lovelace, the women came either individually or in pairs. As a result, their test batteries could be compressed to an intense five days. The schedule was so demanding it sometimes did not allow time for meals. Like Cobb, these women experienced exactly the same tests as the male candidates. In the end, twelve of them (plus Cobb) passed the tests at levels comparable to the best eighteen of the thirty-two male candidates (68 percent for the women versus 56 percent for the men). Those women came to be known as the Mercury Thirteen, but they were never officially astronaut candidates.

The reluctance of NASA to open additional test facilities to the female candidates truncated the overall testing sequence. But, certainly, in the New Mexico phase, the women proved themselves more than capable. After all of the physiological results

Apollo 15 astronauts Scott and Irwin practiced driving an LRV and collecting soil and rock samples near Taos.

were evaluated, Cobb was among the best, her test results putting her in the top 2 percent of all male and female candidates.

Astronauts in Training

New Mexico has provided resources for training space travelers for decades. The Mercury astronauts came to Holloman Air Force Base for zero-gravity experiences before their space flights. NASA's objective was for each astronaut to experience "at least thirty periods of maximum attainable weightlessness in a comprehensive test of mental and physical alertness." The transport plane used for this experience gave the men room to float and maneuver during the 15-second weightless episodes that occurred periodically during the aircraft's series of steep climbs and descents.

After two suborbital flights, the Mercury Program sent one man at a time into Earth orbit for as many as twenty-two trips around the planet. By the time the program ended in 1963, NASA had begun accepting applications for a larger pool of astronauts for the next manned space flight programs. Gemini's ten flights in 1965–1966 developed skills, such as docking with other space vehicles and performing space walks, that would be necessary for the third program. Ultimately, Project Apollo would land men on the moon and return them to Earth.

While on the moon, the astronauts would explore its geology, extract core samples of the ground, and select significant rock and soil specimens to bring back to Earth. To prepare for these tasks, astronauts participated in geology training exercises at several locations around the world where the landscape was somewhat similar to

The Rio Grande Gorge is a rift, or crack, in the Earth's crust that was accompanied by volcanic activity.

101

the conditions they would find on the moon. Between 1964 and 1972, at least seven sites in New Mexico were used for those exercises:

> Philmont Ranch in the northeastern part of the state, where the Southern Rocky Mountains meet the Great Plains
>
> Valles Caldera, a collapsed volcano in the Jemez Mountains in the northern part of the state
>
> Zuni Salt Lake, in a low-profile volcanic crater located in the western part of the state
>
> Kilbourne Hole, a dry, low-rimmed volcanic crater in the southern part of the state
>
> San Juan Mountains, a rugged volcanic range in northern New Mexico
>
> Capulin Mountains, near Raton, with a 1,000-foot-high, extinct, cinder cone volcano
>
> Rio Grande Gorge, near Taos

The final three Apollo missions, in 1971 and 1972, required a different kind of training. On each of those visits to the moon, two astronauts would explore larger areas of the surface by driving around in a Lunar Roving Vehicle (LRV), an electric cart designed to operate in reduced gravity. Astronauts took their driving lessons in northern New Mexico. The lunar landing site for Apollo 15 was near Hadley Rille, a narrow, meandering chasm 1,000–1,300 feet deep. The closest analogy on Earth is the 800-foot-deep Rio Grande Gorge. In preparation for that flight, astronauts David Scott and James Irwin practiced driving the Earth-bound version of the LRV along the rim of the gorge near Taos.

The only professional geologist to visit the moon was Harrison Schmitt, who explored the lunar surface during the Apollo 17 mission in 1972. A native New Mexican, Schmitt was born in Santa Rita and grew up in Silver City. After leaving NASA in 1975, he was elected to the US Senate, representing his home state. He served as the ranking Republican on the Science, Technology, and Space Subcommittee until 1982, when he lost his reelection bid. His successful opponent, Jeff Bingaman, used the campaign slogan "What on Earth has he done for you lately?"

As recently as 1999, astronauts practiced identifying geological features and collecting samples near Taos, in anticipation of more visits to the moon or missions to Mars. Besides learning the characteristics of rocks, they also practiced techniques for locating underground deposits of water.

As NASA continued its Space Shuttle program, Shuttle pilots took most of their flight training at White Sands. The Shuttle reenters the Earth's atmosphere without power, flying as a glider, so pilots cannot practice flying it until their first actual space mission. To get the necessary experience handling the bulky craft, pilots practice landing a specially-designed Shuttle Training Aircraft (STA) that mimics the Shuttle's behavior during a return to Earth. The primary training site is White Sands Space Harbor, an airstrip that is the second backup landing site for returning Space

Shuttles (after the Kennedy Space Center in Florida and Edwards Air Force Base in California).

"The STA is great! It flies very closely to the actual orbiter," said Eileen Collins, who flew the Shuttle Discovery in 2005. "The handling qualities are close enough that we have no problems transitioning from training to the real thing."

Resources

Highest Step in the World, The. Foolish Earthling Productions, DVD, 2003.

Kittinger, Joseph, and Martin Caidin, *The Long Lonely Leap*. New York: E.P. Dutton, 1961.

Kittinger, Joe. "The Sky Is My Office," presented at the National Museum of the Air Force, April 22, 2009. Accessed at http://www.nationalmuseum.af.mil/shared/media/document/AFD-090811-059.mp3.

Mallan, Lloyd. *Men, Rockets and Space Rats*. New York: Julian Messner, 1955.

Meeter, George. *The Holloman Story: Eyewitness Accounts of Space Age Research*. Albuquerque: University of New Mexico Press, 1967.

Swenson, Loyd, et al. *This New Ocean: A History of Project Mercury, NASA, 1989*. Accessed at http://www.hq.nasa.gov/office/pao/History/SP-4201/toc.htm.

Weitekamp, Margaret. *Right Stuff, Wrong Sex: America's First Women in Space Program*. Baltimore: The Johns Hopkins University Press, 2004.

Wolfe, Tom. *The Right Stuff*. New York: Farrar, Strous, Giroux, 1979.

Chapter Six: Window on the Universe

The clear skies and high elevation of New Mexico offer good views of the sun, moon, and stars. From prehistoric times to the present day, people have taken advantage of those views to plan crop planting times, conduct calendar-sensitive religious festivals, investigate the origins of the universe and the nature of heavenly bodies, and evaluate destinations for space travel.

Ancient Astronomers

Astronomers have been building observatories in New Mexico for more than a thousand years. Those early skywatchers did not use telescopes, but they built structures that showed the exact day of events such as the sun's solstices (the shortest and longest days of the year). This was important for the Jornada Mogollon people, who lived in south-central New Mexico and northern Mexico, for example. They wanted to know when winter would begin to fade away and days would gradually grow longer and warmer (the winter solstice). They needed to know when to plant their corn to take advantage of the monsoon rains, which begin just after the summer solstice.

Wizard's Roost and Wally's Dome, a pair of stone observatories in the Sacramento Mountains, show the astronomical skill of these early New Mexicans. Anthropologist Peter Eidenbach was lucky enough to discover Wizard's Roost in 1977. He had been hired to do an archeological survey in connection with a possible expansion of the Ski Apache resort.

"We did the survey and found, almost the last day, a whole series of rock features near the ski area that we couldn't explain," Eidenbach said in a March 2, 2010, radio interview. "It was too high in elevation to be an eagle trap, which was one suggestion we came up with. There was no reason to be up on this ridge point, perfectly exposed, at more than 10,000 feet. And, I guess, our last-ditch explanatory hypothesis was the possibility that, who knows, maybe this is some kind of observatory site."

Eidenbach found several notable features at Wizard's Roost that confirmed this hypothesis. Because of the site's location, the rising and setting positions of the sun at the winter and summer solstices coincides with major peaks in the mountain range. Several rock cairns (manmade piles of stones) at the site appear to point toward several of the brightest stars in the sky. Perhaps most dramatic is the primary rock cairn, which Eidenbach described as "kind of an aboveground rock foxhole—enough space for one person to spend the night." Using his knowledge of history and indigenous cultures, he imagined how the pile of stones with gaps between them was used: "I have often envisioned it as an elder, with assistance from some younger men, going up there at winter solstice, December 20th–21st. It's cold at 10,600 feet, very cold, especially on an exposed ridge point. I'm sure that this feature would have been lined with a buffalo robe, with another robe to go over the top, so that this elder could peek out at dawn and view the alignment and essentially use it as a way to calibrate their calendar."

Nothing has been found at Wizard's Roost that sheds light on when it was built, but scientists believe it probably dates to sometime between 100 BC and 900 AD. Once Eidenbach was convinced that it was truly an observatory, he began to look for other, similar sites in the vicinity. But looking for an astronomically arranged pile of rocks in the Sacramento Mountains is very much like searching for the proverbial needle in a haystack. Although he expected that another observatory might be found on a solstice line of sight from Wizard's Roost, that covered a lot of territory.

Luck struck again in 1979, when Wally Hesse, a prospector who had gotten permission from the Mescalero Apaches to wander around their land, told Eidenbach about an unusual rock formation he had seen. He agreed to take Eidenbach and his associate to see the site. When Eidenbach realized they were driving toward Pajarito Peak, he began to get excited. Viewed from Wizard's Roost, the sun rises behind Pajarito Peak at the winter solstice. Arriving at the site, now called Wally's Dome, they found many rock cairns that align with prominent stars. Most notable was a solar observatory consisting of an arrangement of vertical rocks with a horizontal rock slab resting on them. They are arranged to form a triangular window between the horizontal slab and its supports. Subsequent observations showed that at sunrise on the summer solstice, this window perfectly frames the sun.

Wally's Dome may be a more recent construction than Wizard's Roost. Scientists place its construction between 850 and 1000 AD. Interestingly, another ancient site may be related to these two observatories. The Three Rivers petroglyph site lies along the line of sight from Wizard's Roost toward the sun's rising place at the summer solstice. The number of petroglyphs (images engraved on rocks) at Three Rivers is so large that it is difficult to assign an overall meaning to the site. However, some figures are exactly half-lit by the sun during the winter solstice. Although this is interesting, it does not conclusively identify the site as astronomical.

Southern New Mexico has no monopoly on ancient astronomy. In the far northwest corner of the state, several sites attest to the skywatching prowess of another group of early New Mexicans. Probable ancestors of today's Pueblo peoples, these "ancient ones" were called the Anasazi by the later-arriving Navajos. The Anasazi occupied a large area around the Four Corners (where New Mexico meets Arizona, Colorado, and Utah) from about 400 to 1300 AD. A culturally advanced people, they left several indications that they were also astronomically sophisticated.

The physical center of the Anasazi culture was in Chaco Canyon. Many settlements of various sizes were scattered across the 54-square-mile valley. More than a dozen "great houses" had hundreds of adjacent rooms and stood up to five stories high. The largest of these structures was Pueblo Bonito, a five-story building with 800 rooms. It probably served as a sort of hotel for Anasazi pilgrims who came to Chaco Canyon for important events. The major axes of this U-shaped structure, which was built around 950–1150 AD, align with solar events such as the equinoxes (the two times each year when day and night are of equal length, in March and September).

An observer standing on an upper floor of Pueblo Bonito, or at the top of a carved stone stairway that ends at the crest of the adjacent cliff, could compare the horizontal position of the rising sun to natural high and low points on the ridge to the east. "I have done so, and it works well from the summer solstice to about the end of October, in the sense of having a geologic profile that has obvious markings," astronomer Michael Zeilik wrote in a 1985 *New Mexico Magazine* article. "Even the planting time from mid-April to the summer solstice is easy to note. The horizon calendar fails, however, for one crucial time. From the end of October until the winter solstice, no clearly noticeable feature scores the skyline."

How could the sun priests fill this crucial information gap? A pair of unusual openings in Pueblo Bonito supply an unproven but plausible answer. These narrow openings, which have been described as either windows or doorways, are located in the corners of two rooms. In the incomplete ruins of Pueblo Bonito, only seven corner openings have been found in 650 rooms. In these two particular rooms, the openings face east. In late October, the rising sun begins to shine through the openings, casting a thin shaft of light on the opposite wall. As the days pass, the shaft becomes wider and moves across the wall. On the morning of the winter solstice, the rising sun creates a square block of light that just touches the corner of the wall.

The eastern-most great house at Chaco Canyon is known as Wijiji, a name given to the abandoned structure by the Navajos when they arrived during the fifteenth century. The name was their word for greasewood, a type of shrub that grows in the area.

The Anasazi carved a stone staircase into the side of the mesa that rises near the Wijiji great house. At the top of the stairs is a narrow ledge, and a few steps away, a white sun symbol adorns the mesa wall, probably painted by a Navajo. Walking along the ledge a short distance brings an observer to three boulders. On the closest boulder, Anasazis pecked cross-shaped and spiral petroglyphs, and Navajos later carved double triangle symbols. Standing at this boulder and looking southeast across Chaco Canyon, an observer will see a natural rock pillar. On the morning of the winter solstice, the observer will see the sun rise exactly behind this pillar.

Anasazi sun priests had to not only identify the exact day of a solstice, but they

Several years after Sofaer and others documented the solar effects at the site, the stones' positions shifted somewhat. The sun daggers no longer perform as accurately. To protect the site as much as possible, it is now closed to visitors.

also had to accurately anticipate the day by about two weeks to allow the people to prepare elaborate ceremonies that they would celebrate on the solstice itself. At Wijiji, this could be done by watching the sun rise behind that same pillar from the vantage point marked by the painted sun symbol near the top of the staircase. This alignment occurs sixteen days before the winter solstice.

Perhaps the most dramatic and well-publicized astronomical feature in Chaco Canyon is the "sun dagger" site high on Fajada Butte, a 450-foot-high, vertical-walled outcropping. Near the top, three vertical stone slabs stand next to each other, perpendicular to the butte's wall and leaning so that their top edges touch the wall. Not quite parallel, their inner edges are about 4 inches apart. The slabs are 6–9 feet tall, 2½–3 feet wide, and 8–20 inches thick. Behind them, two spiral petroglyphs are incised into the butte face. The larger spiral is about 15 inches across and consists of 9½ turns. The smaller spiral, with only 2½ turns, is about 4 inches across. Serendipitously, archaeoastronomer Anna Sofaer discovered the significance of the site while she was cataloging rock art in the summer of 1977. With the summer solstice near, she happened to notice that the midday sun, partially blocked by the vertical slabs, cast a narrow wedge of light very nearly through the center of the large spiral.

Subsequent observations revealed that at the summer solstice, the dagger of sunlight cut exactly through the center of the large spiral. At the winter solstice, two vertical sun daggers just touched opposite sides of the large spiral. At each equinox, a shorter sun dagger passed vertically through the center of the small spiral, and a longer dagger passed vertically along the fourth or fifth spiral to the right of center in the larger petroglyph. "The Fajada Butte construct is unique in archaeoastronomy as the only device known to use the passage of the midday sun to create a solar calendar," Sofaer wrote in *Chaco Astronomy.*

Another astronomical site in Chaco Canyon is near the top of the canyon's western wall, above the ruins of a great house called Peñasco Blanco. Three red pictographs are painted on the horizontal surface of a rock overhang. A star with ten flame-like rays radiating from it is beside a crescent moon, and a left handprint is above the moon. "This pictograph is generally accepted as being a depiction of the supernova that created the Crab Nebula," astronomer Dennis Ward wrote in "Sunwatchers: A Survey of Chacoan Astronomical Sites." The exploding star became so bright in July 1054 that it was clearly visible in the daytime for more than three weeks. Calculations show that on the fifth day of that month the relative positions of the supernova and the moon are exactly as indicated in the pictograph. "With the apparent width of the moon being about half a degree, this pictograph comes basically as close as it possibly could to being a true scale rendition of the 1054 supernova seen in conjunction with the waning moon," Ward wrote.

The handprint marks the site as being sacred.

First Photographs from Space

Twentieth-century astronomers watched extraterrestrial bodies with different tools and for different reasons than the prehistoric sungazers had. Beginning in 1858, various people tried to take photographs of the Earth from high above, sending cameras aloft with hot air or gas-filled balloons, kites, pigeons, airplanes, and rockets. The arrival of the German V-2 missiles at White Sands offered the possibility of record-breaking photos.

The first attempt came on October 24, 1946, when a 35-millimeter movie camera was mounted in a V-2 and sent 65 miles into the sky. The rocket fell back to Earth and crashed into the ground at 350 miles an hour, smashing the camera. But Clyde Holliday, the engineer who designed the camera, had anticipated a hard landing and built a steel cassette to protect the film. A 2006 *Air & Space* magazine article quoted a soldier assigned to the retrieval team as he described the scientists' reactions to the film's safe recovery: "They were ecstatic, they were jumping up and down like kids." After the film was developed, the project participants gathered for a screening. The soldier said that "when they first projected [the photos] onto the screen, the scientists just went nuts." The pictures clearly showed the Earth's curvature.

Subsequent V-2 flights filmed the Earth from as high as 100 miles. Holliday wrote in a 1950 *National Geographic* article, "On these photographs we saw what a passenger on a V-2 would see if he could stay alive on the zooming ride up to that height and back again, and how our earth would look to visitors from another planet coming in on a space ship."

Clear for Space Travel?

Some of the New Mexico researchers wanted to see the Earth from space, but others wanted to look the other direction. Astronomer Clyde Tombaugh was one of them. His observations of deep space had been sidetracked by his work with the V-2 program at White Sands, where he developed telescopes and cameras for tracking the missile flights. As that program wound down, he turned his thoughts to a different problem that involved the safety of future space travelers.

As scientists began to seriously consider launching manned spacecraft, astronomers debated the probability of a spaceship being destroyed by colliding with a random space rock. In 1946, one astronomer predicted this fate for one of every twenty-five ships traveling to the moon. In 1951, another astronomer recommended thick shielding on spacecraft to protect against the almost certain impacts. Tombaugh decided to assess the threat by searching for natural debris that might be orbiting the Earth or passing near it, particularly within a distance of 40,000 miles. This debris might have been invisible to telescopes aimed at more distant objects—the moon and beyond.

Funded by the United States Army, Tombaugh undertook this study from 1953 until 1958. In the middle of that time, he left the White Sands Missile Range and became an Earth sciences professor at the New Mexico State University in Las Cruces (he was later instrumental in establishing a separate astronomy department at NMSU). This small shift in location did not affect the study, as telescopic observations were being done in Ecuador. Not only was that location at a high elevation, but it was located at the equator. Tombaugh knew that science predicted orbiting debris would naturally migrate to an equatorial path.

Tombaugh and his staff designed the tracking telescope for the study and evaluated the images it produced. He concluded, in his 1958 final report, that there were no objects near the Earth that posed a threat to spacecraft. The timing of his study was fortuitous, considering the torrent of manmade objects orbiting the Earth that began with the Soviet Union's launch of the first artificial satellite in 1957.

A negative result is not necessarily a negative thing. In this case, it cleared the way for manned space flight. This important information is not part of Tombaugh's public legacy, however. He told biographer David Levy, "I did things that were fully equal to [my discovery of] Pluto. The work out there [at White Sands], the study of Mars, finding the supercluster of galaxies. But all they think of is Pluto. This is a disappointment in that the public did not attach importance to the other things that I did. . . . From the standpoint of real contribution to science, it isn't always the flashy stuff that really counts."

Tombaugh died in 1997, and the New Horizons spacecraft is carrying some of his ashes. It will pass near Pluto in 2015.

Stargazer

Ground-based optical telescopes look at celestial objects through miles of air, which distorts and filters out some of the light, resulting in degraded images. Moving a telescope above the atmosphere results in images that are clearer, more detailed, and more accurate. The success of Project Manhigh suggested the possibility of sending an astronomer into the upper stratosphere in 1959.

"I got a call from the Smithsonian Observatory telling me that they had a guy with a telescope, wanting to go up and would I take him up there," Joe Kittinger told an audience at the National Museum of the Air Force in 2009. "I thought that would be a lot of fun."

The Air Force had a two-person, closed capsule left over from Project High Dive, and they agreed to let Kittinger take astronomer William White above 95 percent of the Earth's atmosphere. Preparing for the flight presented substantial challenges. The occupied capsule would be the heaviest payload ever carried by a balloon, so a stronger balloon had to be designed and built. Technology had to be developed to stabilize the 12½-inch diameter telescope, which would be mounted on top of the capsule. Mechanisms had to be created to control the telescope, which was to be mounted on a turntable. The astronomical observations had to be planned, with the idea of not only getting unobstructed views of certain objects, but also evaluating the atmospheric effects of air turbulence and water vapor.

With all the preparations finally complete, Kittinger and White were sealed in the 7-foot-diameter aluminum capsule at Holloman Air Force Base on the morning of December 13, 1962. The huge balloon lifted the 14-foot-tall capsule (including the roof-mounted telescope) off the ground. Two hours later, the balloon hovered at an altitude of 82,000 feet. For the next thirteen hours, White made his observations while Kittinger assisted him and monitored the operation of the balloon and capsule. After an uneventful descent, they landed near Lordsburg, in the southwest corner of New Mexico, at four o'clock the following morning.

The balloon and capsule systems performed well throughout the flight, but the astronomical observations did not meet expectations. The engineers who designed the telescope stabilization system worked on improving it. The following April, Kittinger and White climbed back into the capsule for a second Stargazer launch from Holloman. Moments before liftoff, a release mechanism malfunctioned, and the balloon took off but left the capsule sitting on the ground. "It was embarrassing," Kittinger said.

"Questions such as . . . 'Are there earth-like planets beyond our solar system?' are being answered by NASA with the help of experiments flown on [unmanned, high-altitude] scientific balloons."—website of Columbia Scientific Balloon Facility, managed since 1987 by New Mexico State University's Physical Science Laboratory in Las Cruces.

Inflating the ungainly Stargazer balloon tested the skills of Holloman's launch crew.

Besides the wasted expense of the aborted flight (the balloon was worth $53,000), the incident raised serious questions for NASA. A static electric charge triggered the balloon's release, raising concerns about the reliability of the switches, which NASA was also using in the Mercury program. Kittinger called his friend John Glenn, who was one of the Mercury astronauts preparing for the first manned rocket flights. Kittinger recalled that Glenn said, "Joe, I'll let you use all the NASA engineers to help you solve the problem, because the one thing we're worried about the most on the Mercury program is we have forty-six different actuators in that program. We worried more about something like happened to you happening to us in that capsule."

About six weeks later, Kittinger and White tried again. Sitting in the pressurized capsule awaiting liftoff, they heard a familiar, small explosion. "The second balloon took off with me sitting on the ground," Kittinger said. "That was the end of the Stargazer program." Funding for the program had run out, and NASA had lost interest in manned, high-altitude balloon flights.

Optical Observatories

Despite their atmospheric constraints, ground-based optical observatories do not suffer from Stargazer's logistical problems. And with its high elevations and frequent clear skies, New Mexico is the home of many significant telescopes, several of which are unusually innovative and powerful.

Perhaps the earliest prominent example of modern times is the National Solar Observatory facility at Sacramento Peak, 40 miles southeast of Alamogordo. It was built in 1951 by the Air Force, turned over to the National Science Foundation in 1976, and expanded several times. The largest instrument at the site is the Dunn Solar Telescope, which was completed in 1969. Its vertical tower rises 136 feet above ground and extends 228 feet under ground. Sunlight enters the top of the tower, travels down to a mirror 188 feet below ground, and is reflected back to an optical laboratory at ground level. Because light coming from the sun is very hot, the air inside the lengthy telescope could become hot enough to distort the light's path. To avoid this, the telescope's interior is kept in a vacuum.

Half a mile south of the Sacramento Peak observatory is the Apache Point Observatory. Like its neighbor, the Apache Point facility includes several telescopes. The largest is 3.5 meters (more than 11 feet) in diameter and became operational in late 1994. It was the world's first large, ground-based telescope to be completely remote controlled. Astronomers located anywhere in the world can control the telescope through their computer keyboards, as long as they have a fast enough Internet connection. This capability not only makes the telescope accessible to a large number of researchers, but it also means that several different astronomers can take turns using the telescope during a single night.

The second largest instrument at Apache Point is a 2.5-meter (more than 8 feet) diameter telescope built in 2000 for the Sloan Digital Sky Survey. The purpose of that project was to create a three-dimensional map of one-fourth of the sky. "The Sky Survey will be the astronomer's field guide to the heavens for the next fifty years," said astrophysicist Bruce Margon, the Survey's scientific director, in a 1998 press release. "Our only current comprehensive guide comes from the Palomar Sky Survey of forty years ago. That survey used photographic plates, making it fifty times less sensitive than the Sloan Digital Sky Survey. It was in only two colors, compared with our five, and of it course it was not in digital form."

Apache Point's elevation of 9,147 feet and its remote location are important to its usefulness. "We're overlooking an Army base, an Air Force base, and a missile range," Russet McMillan, a telescope operator, said in a 2009 webcast called *Around the World in 80 Telescopes*. "All of those large spaces of land are not going to be developed considerably in the future, so we're not expecting a large increase in light pollution."

By the time the first two phases of the Sky Survey were completed in 2008, astronomers had detected 230 million celestial objects and recorded spectra of 930,000 galaxies, 120,000 quasars, and 225,000 stars. Among the discoveries were the most distant quasars in the universe, the dimmest stars and galaxies seen so far, and even a previously unnoticed dwarf galaxy that seems to be merging with our own Milky Way Galaxy.

"That aspect of the unknown—of seeing things that no human being has ever seen before, of going to distances that no one has even really conceived of before—is probably the most exciting aspect of this work," John Barentine, one of the eight telescope observers at Apache Point, said in a *Science Bulletin* from the American Museum of Natural History.

"Huge amounts of data have come from this telescope," Bruce Gillespie, the observatory's operations manager, said in the *Science Bulletin*. "And when I say huge, I'm talking about more data than that contained by the digitized Library of Congress."

In 2005, Los Alamos National Laboratory (LANL) in northern New Mexico became a member of the Sky Survey, which gave it access to all of the data. Physicists at LANL used their extremely powerful supercomputers to search the data for information about dark matter and dark energy. Stars, planets, and interstellar gas are made of ordinary matter, which consists of protons, electrons, and neutrons. Dark matter, on the other hand, is virtually invisible. It never heats up nor cools down, it has no electrical charge, and it can neither emit nor absorb light. Some bits of dark matter consist of ordinary atomic particles such as neutrons, but some are strange particles such as neutrinos. Dark energy is a mysterious force that seems to be accelerating the expansion of the universe.

The LANL researchers found that dark energy makes up 74 percent of the uni-

A third phase of the Sloan Digital Sky Survey began in 2008 and will continue until 2014. "That telescope is moving on to other missions, continuing to probe the structure of the universe, the structure of our galaxy," said Russet McMillan. "They've also had quite a large search in progress, checking 11,000 stars to see if they have planets around them."

verse and dark matter makes up 22 percent of the universe. "This is a very stimulating time for physics," physicist Emil Mottola said in a 2007 article in the LANL science and technology magazine. "Fully 96 percent of the universe seems to be composed of stuff we've never seen directly on Earth!"

LANL has its own observatory, on Fenton Hill in the Jemez Mountains, about 30 miles west of Los Alamos. Several of the instruments located there are RAPid Telescopes for Optical Response (RAPTORs). One of them is paired with a twin located 20 miles away to achieve stereoscopic vision, which allows observers to distinguish between distant objects and space junk orbiting the Earth. It would also allow them to detect a "killer asteroid" heading for Earth.

The RAPTORs are designed to swivel to any point in the sky within a few seconds so they can focus on interesting transient events. They are also robotic telescopes programmed to recognize and focus on interesting events as they happen. For example, in February 2006, one of the telescopes locked on to the afterglow of a 7-second-long gamma-ray burst (caused by a star collapsing to form a black hole). The telescope watched the afterglow for an hour, taking a photograph every 30 seconds. During that time, the telescope's computer analysis system recognized that the afterglow was behaving in an unexpected manner, rebrightening rather than fading. At that point, the RAPTOR placed a cell phone call to one of the researchers, awakening him and alerting him to the unusual event.

"This was a first, an autonomous optical telescope finding an anomaly on its own with no human intervention," said Tom Vestrand, RAPTOR's team leader, in a 2008 article in LANL's science and technology magazine. "If humans had been in the loop they would have said, as we did, 'Gamma-ray bursts don't act like that. Forget it.' And RAPTOR wouldn't have found anything." Instead, astronomers now know that a gamma-ray burst can be followed by an intense glow of visible light. This previously unknown phenomenon may lead to more frequent and better observations that will help scientists understand the mysterious, extremely high-energy gamma-ray bursts that occur throughout the universe.

New Mexico's newest optical observatory is on Magdalena Ridge, about 30 miles west of Socorro. At an elevation of 10,500 feet, it is the world's fourth highest observatory. Its stationary, 8-foot-diameter telescope has been operating since 2006. In 2010, construction began on a movable array of ten 4½-foot-diameter telescopes arranged in a Y pattern. A computer combines the images from the ten component antennas and generates very sharp images of deep space.

A major goal of the Magdalena Ridge Observatory will be to search for "friendly planets," whose atmospheres are chemically similar to the Earth's. In particular, it will look for exoplanets—planets in other solar systems—in the Milky Way Galaxy. "I spent more than half my life waiting for exoplanets to be confirmed, and I've spent the past fifteen years keeping track of them. I have spent my entire life thinking about life in the galaxy," said Dr. Penelope Boston, an astrobiologist working on the project, in a New Mexico Institute of Mining and Technology news release. "One question we have burning within us is, 'Are we alone in the universe?' Up until now, we've been restricted to our own solar system in trying to directly make observations of life processes."

Radio Telescopes

The kind of telescope most people are familiar with is the optical type, which collects and focuses light beams to create a visual image of a distant object. As described above, optical telescopes can be very complex and capable of producing computerized data. Another type of telescope is one that collects and focuses radio waves to produce an image of an object's electromagnetic emissions. This image does not show what the object looks like, but rather its energy behavior patterns.

One of the world's most famous radio telescopes is the Very Large Array (VLA), located on the Plains of San Agustín, about 50 miles west of Socorro, New Mexico. A facility of the National Radio Astronomy Observatory, the VLA is a project of the US government and is available for use, on a time-available basis, by astronomers who submit research proposals. Built in the 1970s, the array consists of twenty-seven dish antennas spaced along the three arms of a Y made of railroad tracks. Each arm of the Y is 13 miles long. The antennas can be moved along the railroad tracks to expand the array to as much as 22 miles across, or contract it to as small as six-tenths of a mile across. Changing the array's overall size changes its resolution. In its largest configuration, it can "see" a small area of the sky in great detail. In its smallest configuration, it can "see" a larger area but with less detail.

Moving the antennas is tricky. Each of them weighs 230 tons and is 94 feet tall, and each dish is 81 feet in diameter. After an antenna is unbolted from its base beside

In its smallest configuration, the VLA's antennas are as close as possible to each other.

> The VLA's railroad tracks were recycled from commercial railroads, including the Atcheson, Topeka, & Santa Fe Railway.

the railroad tracks, a specially designed transporter vehicle moves underneath it and lifts it off the base. The transporter moves slowly down the track to the antenna's next destination and sets it on another base, and workers bolt it securely to the base. The VLA generally operates in one of four configurations and is shifted to the next configuration about every four months.

"When the original planners for the Very Large Array looked around, there were certain criteria that they needed to place this array," said Rick Perley, a VLA astronomer, during the *Around the World in 80 Telescopes* webcast. "They include being at a high elevation to get above as much of the atmosphere as is reasonable, far to the south so we can see more of the sky, and on a very large and flat plain because one of the unique aspects of this telescope is that we can move the antennas around in order to change our resolution."

"The Plains of San Agustín roll flat before you," Michael Zeilik wrote in a 1984 *New Mexico Magazine* article. "For over tens of miles, the grades are no more than 3 percent. . . . Because of the surrounding rim of mountains and the lack of population, little radio noise or signals reach San Agustín. Quiet, high, and dry, it is the perfect place for the [VLA]."

One of the phenomena the VLA astronomers have investigated was a mysterious source of radio waves coming from the very center of the Milky Way Galaxy. The motionless source seemed too small for the amount of radiation it was producing. After studying it for several years, the astronomers realized they were observing a massive black hole. It contains a mass four million times that of our sun, compressed into an area whose diameter is only one-twentieth of the distance from the Earth to the sun.

In another interesting experiment, researchers used the VLA as a receiver for a radar beam that was sent from an observatory in California and bounced off the planet Mercury. Analysis of the reflected beam showed that Mercury has super-cold water ice in the bottoms of craters near its north and south poles. The sun heats exposed surfaces on Mercury to nearly 800 degrees Fahrenheit, but the perpetually shaded crater floors can be as cold as 235 degrees below zero.

Between 1993 and 1997, the VLA collected images that together covered 82 percent of the sky that is visible from its location. Studying those images, astronomers discovered a huge hole in the universe, an area nearly a billion light-years across that is empty of both ordinary matter and dark matter.

In 2009, the VLA recorded the most distant object ever observed, a star that exploded 13 billion years ago, only 700 million years after the estimated time of the Big Bang. As impressive as such observations are, the VLA scientists began to feel their observatory was no longer cutting edge. In 2010, they completed an upgrade—the

Extended Very Large Array (EVLA)—that increased the observatory's sensitivity by a factor of ten. It could detect a signal as weak as a cell phone transmission from as far away as Jupiter. The upgrade consisted of installing new electronic equipment on each antenna, fiber optic cables to carry the received signals to the correlator (a specially designed computer that integrates the signals from all of the antennas), and an entirely new correlator.

In 1993, another radio telescope array was built on an international scale. The ten antennas of the Very Long Baseline Array (VLBA) are scattered over a distance of more than 5,000 miles, ranging from Hawaii to the Virgin Islands. Two of the antennas are located in New Mexico—one in Pie Town, west of the VLA, and one near Los Alamos. The VLBA's control center is in Socorro. The antennas are similar in size to those of the VLA, but together they would have the sensitivity to read a newspaper in Los Angeles while standing in New York.

"The universe is a laboratory unlike any we can build, with tremendous temperatures, pressures, densities and powerful electric and magnetic fields," Dave Finley, public information officer for the National Radio Astronomy Observatory in Socorro, wrote on the NRAO website. "The VLBA is one of our most powerful tools for gaining new scientific knowledge from this natural laboratory."

NRAO is also part of a group of organizations that is building the Long Wavelength Array (LWA), which will ultimately use nearly 13,000 linked antennas to collect very long wavelength signals from space. The LWA will collect radio waves between 13 and 50 feet long. Pyramid-shaped antennas, which are 5 feet tall and 9 feet wide, will be arranged in grids, with 256 antennas clustered in an area the size of two side-by-side football fields. The first of these stations was completed in 2010, and others will be added in subsequent years. The first station, which is already collecting data, is located near the VLA. As many as fifty stations will be scattered across southwestern New Mexico, and perhaps beyond.

Monitoring Mars

Larry Crumpler looks out into space from a different kind of window. Rather than using a ground-based observatory, he remotely "rides" a robotic vehicle as it drives around on Mars. Crumpler, who is the research curator at the New Mexico Museum of Natural History and Science in Albuquerque, is a geologist and expert on volcanoes. He analyzes the photographic images and spectrographs that the rover radios back to Earth, to gather information about Mars' geologic history. He and several other scientists meet for Internet conferences every few days to plan the rover's activities.

The Mars Rover, named Spirit, landed on Earth's neighbor planet in early 2004 after a seven-month journey. Its original mission was planned to last for three months, but Spirit kept responding to signals from its controllers on Earth and kept sending back information. In late 2004, Crumpler and his colleagues drove the rover to the top of a ridge that is now known as Larry's Lookout. The geologist was surprised by what he saw from that vantage point—evidence of sediment-like material. "This all indicates that we're not just looking at volcanic rocks or old broken up rocks, but there is some sort of organized layering," Crumpler wrote in a 2005 article published on the Universe Today website. "Every day there has been something different that we hadn't seen the day before, or some new perspective of the terrain, so I always say that 'today' is the most exciting part of the mission."

After six years of exploring, the rover's wheels got bogged down in soft sand in early 2010. Crumpler and his associates continued to operate it as a stationary base to examine the soil, image the surrounding terrain, and observe the Martian weather. "We are essentially doing a natural history expedition to another world, much like explorers a century or two ago in the American West and other unexplored lands," Crumpler wrote.

Resources

"Apache Point Observatory" and "The Very Large Array" video segments on the *Around the World in 80 Telescopes* webcast, April 3–4, 2009. Accessed at http://www.ustream.tv/recorded/1340142 and http://www.ustream.tv/recorded/1337392, respectively.

Holliday, Clyde T. "Seeing the Earth from 80 Miles Up," *National Geographic*, October 1950, pp 511–528.

Kittinger, Joe. "The Sky Is My Office," presented at the National Museum of the Air Force, April 22, 2009. Audio recording accessed at http://www.nationalmuseum.af.mil/shared/media/document/AFD-090811-059.mp3.

Levy, David H. *Clyde Tombaugh: Discoverer of Planet Pluto*. Tucson: The University of Arizona Press, 1991.

The Michael Shinabery Show, KRSY AM 1230, Alamogordo, March 2, 2010. Accessed at http://www.alamoam.com.

Chapter Seven: Alien Visitors

For as long as people have dreamed of traveling into space, they have also speculated about beings from other planets visiting Earth. Just as New Mexico has provided the open space and sparse population for human space research, it has produced a good share of UFO reports. While the term *UFO* literally means *unidentified flying object*, popular culture usually confers on it the implication of *alien spacecraft*. An ABC News poll conducted at the beginning of the twenty-first century revealed that 25 percent of Americans believe in alien visits, although most of them acknowledge they have seen no proof.

A healthy debate continues between those who believe in the real possibility of incoming space traffic and those who are skeptical, primarily because of the lack of hard evidence. One of the most famous episodes, which still has proponents on both sides of the debate, occurred near Roswell in 1947. It is commonly known as the Roswell Incident.

The Roswell Incident

There are various accounts of the Roswell Incident, but the commonly told elements of the story are that a foreman working on a ranch northwest of Roswell heard a distant explosion during a severe thunderstorm on the night of July 4, 1947. When he rode his horse out the next morning to check the ranch animals and equipment for storm-related damage, he came upon an unusual sight. Scattered over an area approximately three-fourths of a mile long and hundreds of feet wide were a large number of fragments including metallic fabric and thin strips of wood-like plastic with strange properties. The metallic fabric, when crushed into a ball, would open back to its original shape, unwrinkled. The strips of wood-like material would not burn or melt when a flame was applied. Other metallic pieces found at the site were described as extraordinarily strong, resisting deformation even when struck with a sixteen-pound sledge hammer.

The foreman, Mac Brazel (pronounced like *frazzle*), collected several pieces of the debris and, two days later, showed them to the Roswell sheriff. Sheriff Wilcox called

Intelligence Officer Jesse Marcel at the Roswell Army Air Field (RAAF). Brazel took Marcel to the field, where Marcel collected much of the debris. On July 8, RAAF officials issued a press release announcing the recovery of a "flying disk." In the meantime, Marcel took some of the debris to Fort Worth, Texas, and showed it to General Roger Ramey. On July 9, Ramey issued a contradictory press release, and the *Roswell Daily Record* reported that the debris was actually the remnants of a high-altitude weather balloon and its radar targets.

The *Roswell Daily Record's* stories about the incident on July 8 and July 9 carried the headlines "RAAF Captures Flying Saucer on Ranch in Roswell Region" and "Ramey Empties Roswell Saucer," respectively. The Roswell Incident faded. Between 1947 and 1969, the Air Force conducted Projects Sign, Grudge, and Blue Book, which were investigations to determine whether UFOs posed a threat to national security and to scientifically analyze data about UFO sightings. None of those projects even mentioned the Roswell Incident in their reports.

The incident was brought back to life in 1978 when a UFO researcher happened to meet Marcel, who said he thought the Roswell debris was unearthly. In 1980, the *National Enquirer*, a sensationalist tabloid newspaper, picked up the story. Public curiosity and UFO researchers' interest began to grow. In 1989, the television series *Unsolved Mysteries* featured the Roswell Incident and invited witnesses to come forward and tell their stories. In 1991, researchers began to interview first-hand and second-hand witnesses about what they had seen more than forty years earlier. Mac Brazel had died in 1963 and was not available for interviews. Some other witnesses remembered seeing bodies of dead aliens at two other sites in southern New Mexico. Most of them charged that the government was hiding spacecraft wreckage and bodies. A film surfaced that appeared to show an autopsy being performed on an extraterrestrial being. The film caught the attention of Steve Schiff, a US Representative whose district included Roswell, and in 1994 he asked for a federal investigation of the incident.

Roswell Refuted

As custodian of all governmental records relating to the Roswell Incident, the Air Force conducted an exhaustive investigation. It issued a series of reports, culminating in the 1997 book *The Roswell Report: Case Closed*.

A major issue was the nature of the debris Brazel had found. The Air Force concluded that it was, indeed, remnants of a military balloon—but not a weather balloon. In 1945, the United States dropped two atomic bombs on Japan to end World War II. The Soviet Union was rushing to develop its own nuclear weapon capabilities. As the Cold War was beginning, the US government wanted to find a way to detect nuclear bomb tests or ballistic missile launches taking place in the Soviet Union. This could be accomplished by keeping a detector at a very high altitude for an extended period of time. Helium balloons offered that potential.

Between June 1947 and February 1949, more than a hundred high-altitude balloons were launched in Project Mogul. The balloons carried acoustical sensing devices to detect Soviet test explosions. More than half of the balloons were launched from Holloman Air Force Base (formerly Alamogordo Air Base) because of the area's sparse population and the fact that commercial air traffic was prohibited from flying over White Sands. The first launch consisted of a string of twenty-nine rubber weather balloons leading a trail of radar reflectors to enable tracking of the flight from the ground. The entire array was 650 feet long. During its six-hour flight, it reached an altitude of 58,000 feet. The Historical Branch of Holloman AFB reported that the balloon equipment was recovered east of Roswell.

Rubber balloons had a tendency to burst as the helium in them expanded at very high altitudes, so Project Mogul soon began using balloons made of polyethylene like those used in Projects Manhigh and Excelsior. The first launch of this type took place on July 3, 1947. It consisted of a string of ten 7-foot-diameter balloons leading a trail of radar reflectors. The reflectors consisted of three-dimensional, triangular arrays of thin, rigid metal. Holloman's Historical Branch reported that this first flight using the new balloon material was only partially successful. The flight lasted 195 minutes and reached an altitude of 18,500 feet. Recovery was reported as "unsuccessful."

Case Closed concluded that Brazel found the remnants of this balloon array. The materials looked strange to anyone not connected with Project Mogul, which was a highly secret operation (the United States did not want the Soviet Union to know it was monitoring their atomic weapons development). The Air Force could not release

The UFO museum is a few blocks from the Roswell Museum and Art Center, which features a large section devoted to Robert Goddard's rocket research.

information about this secret activity, so it called the object a weather balloon. People who knew what weather balloons looked like dismissed this as a cover-up.

"As early as May 1948, polyethylene balloons coated or laminated with aluminum were flown from Holloman AFB and the surrounding area," *Case Closed* reported. "Beginning in August 1955, large numbers of these balloons were flown as targets in the development of radar guided air to air missiles. Various accounts of the 'Roswell Incident' often described thin, metal-like materials that when wadded into a ball, returned to their original shape. These accounts are consistent with the properties of polyethylene balloons laminated with aluminum."

As for the reports of alien bodies, the Air Force investigation concluded that the "bodies" described by witnesses were related to other programs, Projects High Dive and Excelsior. In those programs, anthropomorphic dummies were dropped from airplanes and balloons to help researchers develop equipment and procedures for parachuting from very high altitudes. The drops did not begin until 1953, but witnesses who claimed to have seen alien bodies were unable to say for sure when they saw them. Furthermore, several witnesses said the bodies looked like dummies or plastic dolls.

Some witnesses recalled seeing hairless bodies being carried on stretchers or laid in wood coffins. The Air Force investigators explained this in *Case Closed*: "The dummies were sometimes transported to and from off range locations in wooden shipping containers, similar to caskets, to prevent damage to fragile instruments mounted in and on the dummy. Also, canvas military stretchers and hospital gurneys were used (a procedure recommended by a dummy manufacturer) to move the dummies in the laboratory or retrieve dummies in the field after a test."

Witnesses recalled that when debris was discovered, Air Force personnel would come out and comb the area, collecting every scrap they could find. Some saw this as an indication of government secrecy. Joe Kittinger had another explanation. "We were directed to remove as much of the material dropped by the balloon as possible," he said in an interview published in *Case Closed*. "Sometimes this was difficult because the balloon and payload would break apart and cover a large area. We collected the debris in these cases by 'fanning out' across a field until we had collected even very small portions of the payload and balloon. We were particularly careful to collect the large plastic balloons because cattle would ingest the material and the ranchers would file claims against the government."

The Air Force investigators documented rational explanations for all the flying saucer and alien sightings involved under the Roswell Incident umbrella. Skeptics thought the government concluding there was no cover-up was like the fox saying he was carefully guarding the hen house from carnivorous invaders.

The Roswell Image

The Roswell Incident continues to have believers and debunkers. Believers are intrigued by the possibilities of alien visitors but cannot produce conclusive, physical proof of their existence. Debunkers propose rational, earthly explanations for the events but cannot prove there has been no cover-up. A *CNN/Time Magazine* poll in 1997, the year *Case Closed* was published, found that nearly two-thirds of the respondents believed an alien craft crashed at Roswell in 1947. More recently, a 2008 poll by Scripps Howard News Service and Ohio University found that a third of adult Americans believe it is either very likely or somewhat likely that aliens from space have visited Earth.

This willingness, perhaps eagerness, to believe the extraterrestrial nature of events such as the Roswell Incident has brought attention and some economic stimulation to New Mexico. The International UFO Museum and Research Center opened in Roswell in 1992. By 2001, more than one million visitors had come to view the exhibits and use the research library. Globes on street lights near the museum are painted to look like alien faces. A McDonald's restaurant down the street is built to look like a flying saucer.

Recognizing the publicity potential, the New Mexico Department of Tourism ran an award-winning advertising campaign in 2007 featuring space aliens visiting the state, with the slogan "The Best Place in the Universe—New Mexico, Earth." The state followed up with a 2008 Tournament of Roses Parade float called "Passport to Our World and Beyond." It featured three aliens modeled after the typical Roswell Incident images—hairless round head that narrowed to a pointed chin, flat facial features, large almond-shaped eyes, and no ears. The float won the Grand Marshal's Trophy for excellence in creative concept and design.

Not Just Roswell

Three sites have generally been included in the so-called Roswell Incident—the debris field discovered by Mac Brazel, a site north of Roswell where a man saw what he thought was a crashed spaceship and several "bodies or dummies," and a site 175 miles northwest of Roswell, in the San Agustín Plains, where people claimed to have seen spacecraft wreckage, dead bodies, and a live alien. Because of the witnesses' vague recollections, the dates of these last two are unknown. However, many other UFO sightings have taken place in New Mexico, some of which were reviewed by Project Blue Book investigators.

After examining 12,618 reported sightings, Project Blue Book listed only about 600 as "unexplained" when it ended operations in early 1970. The list included locations from all over the world, including about 500 in the United States. At least twenty-eight of the unexplained sightings were in New Mexico. Only two states had more—Texas with forty-seven and California with thirty-nine. Both of those states have much larger populations than New Mexico and are larger in area. If aliens from outer space are, in fact, exploring the Earth, perhaps they are drawn to New Mexico for much the same reasons that human space researchers have been—sparse population, high elevation, and large expanses of flat terrain.

Trained Observers and UFOs

Skeptics may wonder about the credibility of those who report seeing UFOs. However, some sightings have been reported by technically adept, trained observers such as scientists, commercial and military pilots, and law enforcement officers. For example, one of Project Blue Book's unexplained phenomena was reported by a group of twelve security inspectors from the Atomic Energy Commission. Shortly before four o'clock on the afternoon of February 25, 1950, the group saw an object in the sky over Los Alamos, New Mexico. They described it as a silver cylinder with tapered ends and flashing lights. Some of the witnesses watched it for as little as three seconds, and others for as long as two minutes, during which time it changed its speed from fast to slow, changed directions, and at times fluttered and oscillated.

Another unexplained sighting took place in April 1949 near Arrey, New Mexico—about halfway between Hatch and Truth or Consequences. In midmorning, an experienced crew was preparing to launch a high-altitude balloon as part of Operation Skyhook, the Navy's secret counterpart to Project Mogul. Before the launch, the crew released a weather balloon to check the wind patterns. As they were watching the weather balloon, they saw an unusual elliptical object in the sky, higher than the weather balloon and traveling in a different direction. It moved rapidly across the sky, then abruptly turned upward and quickly moved out of sight. Charles Moore, one of the crew members, used a theodolite to track the object's movements for about a minute. His measurements indicated that the object was 40 feet wide and 100 feet long, and its altitude was 35–55 miles. Astonishingly, its speed was 5–7 miles per second.

In August 1949, Clyde Tombaugh, his wife, and his mother-in-law shared a UFO sighting from Tombaugh's back yard in Las Cruces. About an hour before midnight, they were enjoying the view of a clear, star-filled sky. In his official statement reporting the incident, Tombaugh wrote, "suddenly I spied a geometrical group of faint bluish-green rectangles of light. . . . The group moved south-southeasterly, the individual rectangles became foreshortened, their space of formation smaller . . . and the intensity duller, fading from view at about 35 degrees above the horizon. Total time of visibility was about three seconds. I was too flabbergasted to count the number of rectangles of light, or to note some other features I wondered about later. There was no sound. I have done thousands of hours of night sky watching, but never saw a sight so strange as this."

This was one of three times Tombaugh saw UFOs. About 1955, he wrote to a UFO researcher named Len Stringfield that "I have seen three objects within the past seven years which defied any explanation of known phenomena, such as Venus, atmospheric optics, meteors, or planes. I am a professional, highly skilled observing astronomer." Tombaugh, like many other observers in New Mexico, saw green fireballs streaking across the sky. "I have seen three green fireballs which were unusual in behavior from scores of normal green fireballs," he wrote to Stringfield. "I think that several reputable scientists are being unscientific in refusing to entertain the possibility of extraterrestrial origin and nature."

National media attention on UFO sightings reached a peak in early 1952, when *Time* and *Life* magazines ran feature stories. "Have We Visitors from Space?", the *Life* article, described ten incidents of credible UFO observations. Fully half of them were in New Mexico. The fact that several of them involved respected scientists enhanced their credibility. The observations of Moore (an engineer who worked on Projects Mogul and Skyhook) and astronomer Tombaugh, described above, are two examples. Another incident described in the *Life* article involved Lincoln La Paz, who was director of the Institute of Meteoritics at the University of New Mexico.

Dr. La Paz asked *Life* not to print his name "for professional reasons," so the article described him only as "one of the U.S.'s top astronomers." However, other sources later identified him as the person who reported the sighting. On July 10, 1947, La Paz was driving in southeastern New Mexico, from Clovis to Clines Corners, along with his wife and two teenage daughters. The day was sunny, but turbulent clouds filled the sky to the west. Shortly before five o'clock in the afternoon, the family saw something strange as they looked ahead toward the clouds. "All four of us almost simultaneously became aware of a curious bright object almost motionless," La Paz told the *Life* reporter. He described the object as having "a sharp and firm regular outline, namely one of a smooth elliptical character much harder and sharper than the edges of the cloudlets. . . . [Its] wobbling motion served to set off the object as a rigid, if not solid body."

They watched the object for more than two and a half minutes as it hovered, disappeared into the clouds, quickly emerged at a much greater altitude, moved slowly across the sky, and finally disappeared into the clouds. Analyzing their observations, La Paz concluded that the object was 160–245 feet long, 65–100 feet thick, and 20–30 miles from the car. He estimated its horizontal speed as 120–180 miles per hour and its vertical speed as 600–900 miles per hour.

La Paz remained objective about the nature of UFOs. Following reports of mysterious yellow disks in the sky near Washington, DC, in 1952, La Paz told an Associated Press reporter that the disks were not mirages or distant planets. "Any suggestion that these come from the depths of space is fantastic," he said. "Their origin is Earth. The question is where on Earth."

Unavailable Information

Sightings by trained, objective observers lent credence to their reports of seemingly unexplainable phenomena. In some cases, information surfaced years later that provided compelling, terrestrial explanations. The case of a UFO report by Lonnie Zamora, a police officer in Socorro, is a good example. A little before six o'clock in the evening of April 24, 1964, Zamora was chasing a speeding car when he was distracted by a roar and a flame in the sky less than a mile away. Concerned that a dynamite shack in that area might have exploded, Zamora hurried to the scene. By the time he got there, the roar and flame had stopped, but he could see a shiny object on the ground.

"It looked, at first, like a car turned upside down," he wrote in his report. "I saw what appeared to be two legs of some type from the object to the ground. . . . The two 'legs' were at the bottom of the object, slanted outwards to the ground. The object might have been about three and a half feet from the ground at that time." Next to the object were two people wearing white coveralls. "These persons appeared normal in shape—but possibly they were small adults or large kids."

He lost sight of the object as he drove closer through the uneven terrain. When he saw it again, the two people were no longer visible. As he watched, the roar began again, and the aluminum-white object began to rise into the air above a column of light blue and orange flame. What he saw and heard so unnerved Zamora that he scrambled away from the rising object, bumping into his car with enough force to knock his glasses off his face. He ran about twenty-five feet and jumped over the top of a hill. "I had planned to continue running down the hill," he wrote. But when the roar stopped, he turned and looked at the rising object. At a height of 10–15 feet, it began to travel horizontally at a fast rate. Then, he wrote, "The object seemed to lift up slowly, and to 'get small' in the distance very fast. . . . It had no flame whatsoever as it was traveling over the ground, and no smoke or noise."

Other officers responded to Zamora's call for assistance. Where the object had stood, they found smoldering brush and grass. The ground showed three shallow, round indentations and four freshly dug, rectangular troughs about 6 inches wide, 16 inches long, and 2 inches deep.

Investigators were unable to come up with a satisfactory explanation for what

The last remaining PEPP aeroshell is on display at the White Sands Missile Range museum.

Zamora saw. In 1975, Lieutenant Colonel Hector Quintanilla, who was head of Project Blue Book from 1963 until 1969, described his work on the project in the book *UFOs, an Air Force Dilemma*. Referring to the Socorro incident, he wrote, "I've always had some doubt about this case, even though it is the best documented case on record. In spite of the fact that I conducted the most thorough investigation that was humanly possible, the vehicle or stimulus that scared Zamora to the point of panic has never been found."

A plausible explanation did not surface until 1995. That is when Duke Gildenberg learned about special tests that had been conducted at White Sands around the time of Zamora's experience. The equipment being tested was a Surveyor lunar lander, an unmanned craft designed to land on the moon. In addition to filming the lunar surface with its television camera, the craft could use a remote-controlled shovel to dig a shallow trench and reveal the appearance of the lunar soil beneath the surface. Surveyor had three slanting legs, which ended in circular pads that matched the spacing of the circular indentations left at the Socorro site. The Socorro troughs were consistent with Surveyor's scooping shovels. Surveyor had three small rocket engines that enabled the craft to make in-flight course adjustments and slowed its descent for a soft landing. Test firing those engines could explain the smoldering vegetation Zamora's fellow officers saw.

Records show a Surveyor test flight scheduled for April 24, 1964. During the test, the lunar landing craft was attached to the side of a small, aluminum-white helicopter. The two-man helicopter crew wore white coveralls.

Although the Surveyor test presents a plausible explanation for the mysterious event, it has not been conclusively proven to be what Zamora saw. Some people believe, instead, that the whole thing was a prank staged by students from the local college. Others maintain that it was, indeed, an alien spacecraft.

PEPP Aeroshell

The most flying saucer-like object sent aloft over southeastern New Mexico in the years following the Socorro incident did not generate any UFO reports. Eight of the fifteen-foot-diameter PEPP (planetary entry parachute program) aeroshells were launched in 1966, 1967, and 1972, some by rocket and some by high-altitude balloon.

Strangely, the other-worldly vehicle was not even an object being evaluated for interplanetary travel. Rather, it was a carrier designed to test various parachute systems that might be used for a soft landing in the thin atmosphere of Mars or Venus. After the rocket or balloon had reached an altitude of about 100,000 feet, the aeroshell's twelve rockets boosted an experimental parachute system to 140,000–160,000 feet at a high speed. This simulated the speed and air resistance that an unmanned craft would experience during a landing on one of Earth's sister planets. Then the aeroshell was jettisoned to crash to the ground, while the payload and parachute system it had carried drifted gently down to a soft landing.

To take advantage of the instrumentation at the White Sands Missile Range that could track the parachute and payload's descent, the aeroshells were launched from the Roswell Industrial Air Center (formerly the Roswell Army Air Field). "In appearance the Viking and Voyager probes could be mistaken for a flying saucer," Duke Gildenberg said in his interview published in *The Roswell Report: Case Closed*. "They were both unclassified highly publicized projects and I do not recall getting any UFO reports for these flights."

Alien Cowboys

Man first set foot on the moon in 1969, but during the 1970s an earthly (and earthy) phenomenon captured the attention of people across the United States and the federal government. Throughout the decade, reports poured in from Tennessee to Oregon that thousands of dead animals (mostly cattle, and some horses) were being found in puzzling circumstances. With little or no blood being spilled, their soft tissues—eyes, tongues, lips, genitalia, internal organs—had been removed with apparently surgical precision. In New Mexico alone, sixty-five cattle and six horse "mutilations" were reported between 1975 and early 1979. The problem seemed to be escalating—forty-nine of those seventy-one cases were reported in 1978.

To the general public, the statistics were intriguing, but to Western ranchers they were bottom line issues. The Gomez family, who raised cattle near Dulce, New Mexico, was particularly vocal and, perhaps, particularly hard hit. During the second half of the decade, they lost at least seventeen cattle to the phenomenon, and the financial consequences drove them out of business.

What was going on? Letters poured in to federal congressmen, and a US Senator from New Mexico responded. Harrison Schmitt, the last—and only scientist—astronaut to walk on the moon, asked the FBI to investigate the bizarre, economically harmful happenings. The agency had no authority to investigate isolated happenings on privately owned ranches, but fifteen of the mutilations had occurred on Indian land in New Mexico—seven on Santa Clara Pueblo and eight on the Jicarilla Apache reservation near Dulce. With jurisdiction confirmed, the FBI opened an investigation into those fifteen incidents, based on information collected at the time of the discoveries.

Around the country, interest in unexplained phenomena was still high. An unusually large number of strange sightings were reported in the Dulce area. Gabe Valdez, a New Mexico state police officer who reported that he saw landing tracks near the dead animals, said he saw unidentified objects in the sky every other night. "FBI Joins Investigation of Animal Mutilations Linked to UFOs," screamed a *National Enquirer* headline in mid-1979. Ultimately, the FBI investigators were stymied. They found no logical explanation for the mutilated carcasses, and closed the investigation.

At about the time the FBI investigation was winding down, an Arkansas sheriff responded to a rancher's report of two mutilated calves. The rancher donated a cow so the sheriff's department could conduct an experiment. The cow was killed by an overdose of tranquilizers and its carcass was left in a field. Observers watched the animal all night and into the next day. Eighteen hours after its death, flies and buzzards had eaten most of its exposed soft tissue and internal organs, leaving no pools of blood. As the day went on, severe bloating ripped the animal's skin apart in several places. Photographs confirmed that the skin tears and the edges of the missing soft tissues looked the same as similar wounds on mutilated animals examined previously. This evidence, however, failed to convince those who believed the mutilations were the work of aliens, Satanists, or covert government operatives.

Dulce Doings

Not only was Dulce at the epicenter of the FBI investigation, but local residents often reported UFO sightings that coincided with discoveries of mutilated animals. A May 1980 incident near Cimarron, New Mexico, 175 miles to the east, would soon bring greater notoriety to Dulce.

Myrna Hansen was driving with her young son when they saw two strange aircraft. UFO investigator Paul Bennewitz, who operated an Albuquerque company that produced high-altitude testing equipment for the Air Force, interviewed Hansen. Then he arranged for a fellow UFO investigator, who was a psychologist, to hypnotize her and recover her memories of the experience. During the next three months, she described a harrowing adventure. She recalled seeing the UFOs carrying cattle away. Then she and her son were discovered and whisked away as well. Alien creatures took them to an underground compound, where she saw cattle being mutilated. The aliens subjected Hansen to a physical examination, and implanted small, metallic objects in her and in her son.

During his investigation of the abduction story, Bennewitz began to receive strange messages and visual images on his computer. He believed they were transmitted to him from the underground alien base. He learned the geographical coordinates of the base's location, which placed it underneath Archuleta Mesa, near Dulce. While visiting the area, he saw large UFOs, even photographing one as it entered the mesa. He said his investigation showed that the underground base was a scientific research laboratory jointly operated by the US military and extraterrestrial aliens.

Bennewitz's claims were widely rejected, but some believers perpetuated the story. Dulce Base became an important outpost in the government/alien conspiracy network. Reportedly, deep tunnels link it with other underground bases in New Mexico (including Sunspot, Carlsbad, and Los Alamos) and beyond—Colorado Springs and even Nevada's notorious Area 51.

The story had become so deeply entrenched that UFO researcher Norio Hayakawa organized the "Dulce Base: Fact or Fiction?" conference in March 2009. The conference, which was held in Dulce, attracted more than 100 attendees from as far away as Pennsylvania and Hawaii. Speakers included local and state police officers, several UFO researchers, and a member of the Gomez family who had lost their ranch because of cattle mutilations. The speakers presented a range of opinions. Some espoused the alien connection, but others claimed a secret base was being operated by the government with no alien involvement. Still others believed the happenings have been created by the government to divert attention from covert activities taking place in the surrounding area.

With no definitive proof, the debate continues.

Trementina Base

There may or may not be an underground facility near Dulce, but there certainly is one near Trementina, 45 miles east of Las Vegas, New Mexico. A highly secure vault, built by the Church of Spiritual Technology, houses archives of the writings of Scientology founder L. Ron Hubbard.

Preserving the writings of prominent people is not unusual, but the Scientologists have gone beyond normal measures. After touring the facility in 2005, Las Vegas (New Mexico) Police Chief Tim Gallegos told the *Albuquerque Journal* that Hubbard's books were reproduced in five different formats including engraved titanium or stainless steel plates and special paper pressure-sealed in titanium boxes. "It appeared to me that they were creating a time capsule," he said.

San Miguel County Sheriff Chris Najar, who took the same tour, told the *Journal*, "When they talk about preservation, they're not talking fifty years. They're talking 1,000 years."

Media interest in the facility was aroused, not by these extreme preservation methods, but by a large logo marked in the ground near the vault. The logo is that of the Church of Spiritual Technology, but the reason for its enormous size is unclear.

A *Washington Post* reporter interviewed people who had once belonged to the secretive Church of Scientology for a November 2005 article about the large logo. "Former Scientologists familiar with Hubbard's teachings on reincarnation say the symbol marks a 'return point' so loyal staff members know where they can find the founder's works when they travel here in the future from other places in the universe," the reporter concluded.

Michael Pattinson, a former Scientologist, appeared on the CNN program *360°* in December 2005. He told the host, Anderson Cooper, that the circles mark the location for reincarnated Scientologists coming from outer space. "I think they're not designed to be seen by human beings, but by alien beings," Pattinson said.

New Mexico and Extraterrestrial Phenomena

Along with its lengthy involvement with space-related research, New Mexico seems to have an unusually high level of perceived UFO activity. Peter White, a folklore expert at the University of New Mexico, suggested an explanation to an *Albuquerque Journal* reporter in 1999. He said, "I think that if belief in UFOs and aliens is a search for spiritual meaning and significance, or spiritual reality, it's very typical and it's very natural and logical that that would happen in New Mexico, because the history of New Mexico has been this search for spiritual values."

The logo at Trementina Base dwarfs the road beside it.

Resources

Contributions of Balloon Operations to Research and Development at the Air Force Missile Development Center. Holloman Air Force Base, New Mexico, 1947–1958. Historical Branch, Holloman Air Force Base, 1959.

McAndrew, James. *The Roswell Report: Case Closed.* Washington, DC: US Government Printing Office, 1997.

Quintanilla, Hector. *UFOs, an Air Force Dilemma.* National Institute for Discovery Science, 2001. Accessed at www.ufocasebook.com/pdf/afdilemma.pdf.

Thomas, David E. "A Different Angle on the Socorro UFO of 1964," New Mexicans for Science and Reason, 2001. Accessed at http://www.nmsr.org/socorro.htm.

Chapter Eight: Xpediting Commercialization

NASA, once the standard bearer of American exceptionalism and the inspiration for young people aspiring to careers in science and engineering, has lost its luster. It no longer sends astronauts on missions to explore extraterrestrial destinations. Several of its unmanned vehicles are sending back information about the moon, Mars, and more distant objects, but they lack the sense of drama and adventure of human explorers. Since 1972, all that has remained of NASA's manned space program is routine shuttling of people and equipment to a single destination, the International Space Station, in low-Earth orbit less than 300 miles above the planet.

Beginning about 1990, a paradigm shift began gathering momentum, led by visionary entrepreneurs and encouraged by NASA. The development of space vehicles moved toward becoming a commercial venture rather than a government operation. Certainly, NASA had a history of contracting with companies such as Lockheed Martin and Boeing to build spacecraft for them, but under the new paradigm, NASA would be only one of many customers for private vehicle designers and manufacturers. In an even greater departure from the past, launches of space vehicles also moved into the commercial realm. "I see a day in the not very distant future, where instead of NASA buying a vehicle, we buy a ticket for our astronauts to ride to low-Earth orbit, or [arrange] for a cargo delivery to space station by a private operator," NASA Administrator Michael Griffin said in a 2008 interview.

The movement to commercialize space exploration sprang up in various parts of the country. A handful of business people, having diverse backgrounds but a common fascination with space travel, ultimately converged on southern New Mexico. By the time they arrived in the early 2000s, they found a framework well prepared by local visionaries.

Grassroots in the Desert

With its long history of space research, the White Sands Missile Range (WSMR) was a natural magnet for increased development. In fact, WSMR was at the forefront of the commercialization movement. On March 29, 1989, a rocket provided by Houston-based Space Services was launched from White Sands, carrying the country's first licensed commercial payload on a suborbital flight lasting 15 minutes and reaching an altitude of 200 miles.

Task Force

Recognizing the value of the local resources and experience, as well as the economic potential for developing a facility near White Sands for commercial space launches, a group of New Mexico business people joined together in 1993 to form the Southwest Spaceport Task Force. Some, like Lou Gomez, had space-industry careers; he had worked at NASA's White Sands Test Facility since the 1960s. Others, like Gary Whitehead, whose Las Cruces automobile dealership was previously owned by retired astronaut Frank Borman, worked in an unrelated business that would reap the general benefits of an upsurge of economic development.

The following year, the state legislature established the New Mexico Office of Space Commercialization within the Economic Development Department. Credibility in the somewhat fanciful objective increased in 1996 when NASA sought bids from companies to design and test a prototype for a reusable launch vehicle designated the X-33. It would be the precursor for a vehicle that could carry payloads to low-Earth orbit. Denver-based Lockheed Martin won the contract and began developing a vertical takeoff, horizontal landing vehicle called VentureStar. As the program progressed, the company requested bids for a test launch and landing site in 1998.

By this time, Lou Gomez was a program manager in New Mexico's Office of Space Commercialization, and he helped put the state's proposal together. Eighteen states bid on the contract, and evaluations and negotiations took more than two years. In the meantime, Lockheed Martin was experiencing delays and technological difficulties with their prototype vehicle. Just as the company was poised to announce its site selection in 2001, NASA canceled the program. "We came out on top, but then NASA pulled the funding," Gomez said.

Winning the competition for a canceled project was a hollow victory, but it validated the potential for a southern New Mexico spaceport. During the following year's political campaign season, people in the area around the proposed site were still hopeful for the economic boost such a facility would provide.

"The first that I had heard the word *spaceport* in New Mexico was campaigning with the candidate for governor, Bill Richardson, in 2002 throughout southern New

> "In the last 40 years, we've never fulfilled the promise that we had seen in Apollo. So, now people are saying, 'I'm going to give up on the government, I'm going to do it myself.'"—Peter Diamandis

Mexico," said Rick Homans, a former Albuquerque publisher, in a 2010 interview. "We didn't really understand it all too well, but we heard from people a lot of excitement, support, and enthusiasm about it." During the campaign, they even included a question about a spaceport in a poll and found that a vast majority of southern New Mexicans supported the idea.

Bill Richardson, a former US Representative and national Secretary of Energy, won the election and appointed Homans to be New Mexico's Secretary of Economic Development. Homans was barely settled in his new office in January 2003 before he had a chance to learn more about the proposed spaceport. Half a dozen members of the Southwest Spaceport Task Force—a group Homans likes to call the *space pioneers*—paid him a visit. Armed with a decade's worth of studying and planning for building a spaceport, the task force members described the geographic, demographic, climatic, and historic facts that made the site right. But they tempered their enthusiasm and commitment with practicality.

"They basically laid out this concept of an inland spaceport that would be very useful and workable with the development of reusable launch vehicles and reusable booster systems and things like that," Homans said. "They said, in a very visionary kind of way, we're not asking you to do anything except to listen and to understand and to wait for the right time. . . . We have to wait for this industry to begin to emerge, and that's when we go forward."

Intrigued but not ready to act, Homans set the idea aside.

The X Prize

Throughout the 1990s, while the Task Force members were pursuing their vision, other people around the country were also exploring ways to get in on the ground floor of the private-sector space industry. Aerospace entrepreneur Peter Diamandis was one of them. His idea not only got the nation's entrepreneurial juices flowing, but it generated the spark that began moving the New Mexico spaceport from dream to reality.

In 1994 Diamandis, a student pilot, read *The Spirit of St. Louis*, Charles Lindbergh's autobiography. He discovered that Lindbergh undertook his historic, thirty-hour, non-stop flight across the Atlantic Ocean in May 1927 to win a $25,000 prize. Raymond Orteig, who owned two successful hotels in New York, offered the prize in an effort to encourage the development of aviation. Nine teams spent a total of $400,000 trying to win the competition, and it took eight years for someone to achieve the required non-stop flight between New York and Paris. The consequences of Lindbergh's success were astounding: in 1927 alone, applications for pilot's licenses tripled, and the number of licensed aircraft quadrupled. The number of US airline passengers in 1929 (173,405) was thirty times what it had been in 1926.

Diamandis became excited about the potential for a similar competition to design, build, and fly affordable, reusable, passenger-carrying space vehicles. After careful consideration, he decided $10 million would be an appropriate amount to offer.

He did not have that kind of money to contribute, so he would have to find one or more financial backers.

What should he call the prize, when he had no generous donor to name it after? He decided to use X Prize, an anonymous, placeholder name. Then two other factors came together in a karma-like way. In the United States, the letter X typically designates an experimental aircraft. And X, in Roman numerals, means ten—nicely representing the X Prize's $10 million award. Diamandis is quoted in the book *Rocketeers* as later recalling, "It was one of those things [where] you reach across time and space and touch a powerful idea that you know is going to work."

The next item on Diamandis' agenda was to define the competition's rules. Exactly what did it mean for the winning vehicle to travel into space? The US Air Force, somewhat arbitrarily, considers flights above 50 miles to qualify for astronaut's wings. Also, somewhat arbitrarily, the Fédération Aéronautique Internationale, the world governing body for air sports and aeronautical world records, considers space to begin at 100 kilometers (62 miles). To eliminate any doubts about the validity of the X Prize winner's accomplishment, Diamandis decided to use the internationally recognized definition. To claim the prize, a competitor would have to build a manned spacecraft without any government funding, return safely to Earth after carrying three people (or a pilot and weight equivalent to two passengers) into space, and repeat the feat in the same vehicle within two weeks.

With the competition details worked out, Diamandis just had to find the $10 million. After obtaining important but insufficient funds from a number of individuals, he saw an intriguing opportunity. Sponsors of big-prize competitions—such a golfer getting a single try to hit a hole in one, or a basketball fan trying to shoot a free throw from half court—do not usually fund the entire prize themselves. Instead, they purchase an insurance policy at a reduced rate that takes into account the very small probability that someone will actually accomplish such a difficult feat. Diamandis approached representatives of the aerospace insurance industry to see if they would write a so-called hole-in-one policy for the X Prize competition. "The insurance industry went to the traditional players—Boeing and Lockheed and so forth—and said, 'Are you going to fly, are you going to compete in this?' and they said 'no,' and they concluded that no one could win," Diamandis said in a 2005 radio interview. "I paid a multi-million-dollar premium that was funded by the Ansari family."

In May 1996, Diamandis announced the X Prize competition. During the next several years, twenty-six teams from seven countries spent a total of $100 million trying to win the $10 million prize. On September 29, 2004, SpaceShipOne carried

Anousheh Ansari and her brother-in-law, Amir Ansari are electrical engineers who immigrated to the United States from Iran. They made a fortune by founding and eventually selling a successful telecom firm. In 2006, Anousheh became the world's first female private space explorer when she flew to the International Space Station on a Russian Soyuz rocket.

a pilot and weight equivalent to two passengers to an altitude of 64 miles (112 kilometers), taking off and landing at the Mojave Airport in California. Five days later, the same craft flew to an altitude of 69.6 miles, less than two months before the X Prize insurance policy would expire. Scaled Composites, a company owned by aeronautical engineer Burt Rutan and backed by billionaire Paul Allen, won the X Prize with an 80-minute flight during which a mothership (WhiteKnightOne or WK1) carrying SpaceShipOne (SS1) took an hour to spiral to an altitude of nearly 10 miles. Then SS1 was released. It fell for a few seconds before its pilot fired the rocket engine, blasting the ship into space. Finally, the spaceship glided back for a landing much like NASA's Space Shuttles make.

Rutan took the novel approach of air launching his spacecraft for reasons of safety and efficiency. Climbing slowly to the launch altitude on a mothership powered by standard jet engines meant SS1 had to fire its rocket engine for only a short time to accelerate in the thin upper atmosphere. A ground launch, on the other hand, would require a much more powerful rocket engine to fire for a longer period, burning a larger amount of more volatile fuel to push a spaceship through the dense, low-altitude atmosphere. That would place large tanks of dangerous fuel near the passengers, increasing the risk of an explosion. It would also require a heavier, sturdier ship to withstand the force of pushing through the dense air. Furthermore, if SS1's rocket malfunctioned when it dropped from WK1, the pilot could simply glide the ship gently back to the landing strip.

Brian Binnie, the pilot of the October 4, 2004, flight, climbed out and told the large crowd of spectators, "It's a fantastic view, there's a fantastic feeling, there's a freedom there and a sense of wonder that, I tell you what, you all need to experience."

Rick Homans had been in a similar crowd three months earlier, when SpaceShipOne made its first full test flight. Pilot Mike Melvill flew it to an altitude of 62.2 miles, only 1,000 feet above the threshold of space. "It was very emotional," Homans said in a June 2004 *Albuquerque Journal* article. "All of a sudden, the rockets started to fire and the ship just went straight up in the sky. You could see the contrail very defined behind it as it went higher and higher, and the crowd just erupted in cheers." He had come to believe in the real possibility of commercial space flights. "It also did not escape us, the huge significance of this morning being the beginning of a whole new industry in which New Mexico will play a very big role," he said.

"I think that Rutan winning is actually one of the very best things that could have happened to the [aerospace] insurance industry," Diamandis told a radio audience in August 2005. He explained that the industry currently had only a dozen or so events to cover each year. "They far more would rather have hundreds or thousands of flights that they can use, like any auto insurance or home insurance, to spread their bet over," he said. "And that's only going to happen when you have the personal space flight revolution won."

X Prize Cup

This confidence was not an overnight revelation for Homans. A few months after his conversation with the Southwest Space Task Force members in January 2003, Homans received an interesting packet in the mail. It was an invitation for New Mexico to apply as a host site for the X Prize Cup, a new brainchild of Peter Diamandis. The event, which would feature competitions for privately developed spaceships, would be an incentive for companies like those that competed for the X Prize to keep working on their vehicle concepts. Diamandis wanted a permanent location for the annual event.

This looked like the catalyst the Task Force members and Homans were waiting for. "As I opened this envelope and laid it out, there were, I think, twenty-four 8-by-10 color glossies of different kinds of technologies from all over the world for reusable launch vehicles," Homans recalled in 2010. "And as I had it out there and saw this description of the X Prize Cup, it just hit me that this is what the beginning of a new industry looks like." He talked it over with his staff and with Governor Richardson. "We realized if we are going to claim the right to help give birth to this new industry, we have to win the right to hold the X Prize Cup," he said. "We don't know what the future of it is, but if we don't grab this right now from the big guys, then we will have lost before the race even started."

Homans was well aware of a huge opportunity for economic growth that the state had lost in the 1970s. Bill Gates and Paul Allen were living in Albuquerque when they founded the software company Microsoft. Potential investors in the community declined to buy stock in the company, and local banks denied its loan applications. Gates and Allen found a more supportive atmosphere in Seattle and moved there. By the 2000s, Microsoft, with Gates still in charge, employed 35,000 people in and around Seattle. "What Gates left behind in New Mexico was a profound lesson to us," Homans told the author of *Destination Space*. "He has always been very kind and doesn't point the finger at his rejection in New Mexico, but I feel very strongly that if New Mexico had developed a different business culture—a culture which embraced entrepreneurs and new technology—then Bill Gates might have stayed."

Determined not to repeat such a loss of opportunity, Homans and Governor Richardson convinced the state legislature to appropriate $9 million to support hosting the X Prize Cup. Lou Gomez was involved again, helping Office of Space Commercialization Director Peter Mitchell prepare the state's proposal. Along with the legislature's financial commitment, the application highlighted the wide-open spaces, sparse population, and elevation of the proposed site (a feature allowing a spacecraft to start almost a mile higher than at a sea-level launch site and reach space using less fuel). An enthusiastic base of support for the spaceport showed New Mexico's continuing commitment to hosting the event for the long term. The facility would not be built in time for the first few years of Cup competitions, but alternate sites existed nearby.

The X Prize Foundation's selection committee evaluated New Mexico's application along with those of California, Florida, and Oklahoma. The winner, announced in May 2004, was New Mexico.

Countdown to the X Prize Cup

"My vision is to try and really jump-start a new industry, which is private or personal spaceflight," Diamandis said in the 2005 radio interview. "The X Prize Cup is our mechanism for keeping that competition going and really bringing about the birth of an industry that's competitive in nature, much the same way you've got Sony and Panasonic and Dell and Apple competing in the free marketplace. That's what is going to bring the price down and reliability up . . . in space flight as well."

Interested companies would need time to develop their vehicles enough to compete in X Prize Cup events, and the organizing committee would need time to define the events and raise the prize money. In the meantime, Diamandis decided to hold a preliminary event to generate publicity and get the public involved. Called the Countdown to the X Prize Cup, it took place in Las Cruces on October 6–9, 2005, hosted by the newly created New Mexico Spaceport Authority. The first day was devoted to an International Symposium on Personal Spaceflight. On the second day, the New Mexico Museum of Space History in Alamogordo invited the public to hear special speakers and view the exhibits. On the third day, private space vehicles were displayed at the Las Cruces International Airport, with static engine firings and flight demonstrations.

"We're bringing . . . these vehicles to New Mexico to try and wow the audiences and give people a chance to touch, feel, see these vehicles, meet the pilots, see rocket engine firings, test flights, and really be part of this new personal spaceflight revolution," Diamandis said. Between 10,000 and 20,000 people attended the weekend event.

In the third day's first dramatic event, former NASA astronaut Rick Searfoss flew XCOR Aerospace's EZ-Rocket on two demonstration flights. The rocket-powered airplane was the first generation of a ship being prepared to compete in proposed Rocket Racing League events. After the first flight, Searfoss told a reporter, "Let me just tell you, it's a kick in the pants."

Armadillo Aerospace, headquartered in Rockwall, Texas, put on the next rocket flight demonstration. John Carmack, who had become famous by creating the cutting-edge video games *Doom* and *Quake*, founded Armadillo in 2000 and registered to compete for the X Prize. After five years of development, he was ready to showcase his vehicle. Using the same type of joystick controller designed for playing video games, Carmack guided the cone-shaped rocket up off the ground. It rose 20 feet into the air and hovered for a few seconds. Then Carmack guided it back down toward the ground.

"Unfortunately, when the vehicle came down for landing, one leg landed on the steel plate we launched from, the other three legs landed off in the mud, and the vehicle tipped over," Carmack wrote a few days later. He and his crew stood the vehicle back upright on the launch pad. It had flown well and appeared to be undamaged, so he decided to fly it again. This time, he planned to fly it 100 feet above the ground. When the crew tried to refill the fuel and propellant tanks, however, they discovered that an interior hose had been punctured during the fall. "This was very disappointing," Carmack wrote. "We brought the vehicle back in to the crowd and pulled the shell off to let people see the insides of the rocket that had just flown, which turned out to be a pretty big hit."

For the third live demonstration of the day, the British company Starchaser Industries conducted a static test of its 5,000-pound-thrust rocket engine. For the Starchaser crew, it may have seemed almost routine; they had previously fired this reusable engine eight times. The audience, though, was anticipating an event they rarely, if ever, had experienced—the thrill of noise, vibration, and flame plume that showed the raw power of a large rocket engine. They got more than they expected. Immediately after the engine ignited, it exploded into a 30-foot fireball. Starchaser's CEO Steve Bennett, who was narrating the test, lightheartedly told the crowd, "We wanted to do a grand finale for the X Prize Cup, so we thought we'd blow our engine up." Firefighters were on hand to quickly extinguish the fire. Afterward, Bennett told a reporter, "Eventually it was going to go *pop*. In the world of rocket science you can learn a lot from an anomaly like this."

In fact, as later reported on the Starchaser website, traces of the fuel (kerosene) moved into the outlet for the oxidizer (liquid oxygen) chamber, resulting in an uncontrolled burn. The fireball and plume of black smoke could be seen for miles, but the engine suffered negligible damage.

2006 X Prize Cup

Peter Diamandis was committed to the idea of offering monetary prizes for spaceship development competitions. "If a company is looking to build a vehicle, and there's ten, fifteen million dollars of prizes on the table that they can go after, fantastic," he said in the 2005 radio interview. "It helps them get sponsorships, which is, frankly, sometimes a lot easier to get than investments, much the same way that Indy car race leagues or Champ Car or NASCAR gets sponsorships." But he also said that when he was running the X Prize program, raising the prize money was, by far, the most difficult aspect of the project. Now he needed to raise money again, to fund the X Prize Cup competitions.

NASA gave him a jump start. As part of its Centennial Challenge program, the federal space agency was offering a total of $2 million in prizes for private development of a lunar lander. In general terms, the rocket-powered vehicle would have to fly a simulated moon maneuver by taking off from a starting place, traveling horizontally, landing in a different place, then taking off again, and returning to its starting place. This would be the first step in ultimately producing a vehicle that could ferry cargo or passengers between a lunar orbiter and the moon's surface. In May 2006, NASA contracted with the X Prize Foundation to manage the competition.

Diamandis now had a funded contest to feature at the first competitive X Prize Cup event, which was scheduled to take place in Las Cruces on Friday and Saturday, October 20 and 21, 2006. Interested teams had less than six months to design, build, and test a vehicle. Four groups accepted that daunting challenge and registered to compete: Acuity Technologies, Armadillo Aerospace, Masten Space Systems, and Micro-Space.

By opening day of the 2006 X Prize Cup, only one of those teams was ready to fly its vehicle in the competition. Two of the teams decided their vehicles were not ready to compete. The third team's vehicle was ready, but it was not allowed to fly because it had not yet qualified for an experimental permit from the Federal Aviation Administration's Office of Commercial Space Transportation (AST). Although those three teams did not compete, they did display their vehicles. Armadillo Aerospace, the only team that was fully ready to fly, had barely managed to qualify in time, completing the AST requirements by flying a brief demonstration flight the day before.

The Lunar Lander Challenge consisted of two difficulty levels, with first and second prizes to be awarded for each level. To win Level One, the vehicle would have to rise to a height of 50 meters (165 feet), move horizontally 100 meters (330 feet), and land on a 10-meter (33-foot) diameter concrete pad. The flight had to last at least 90 seconds. After being refueled, the vehicle would have to repeat the flight with the same restrictions, returning to its original launch pad. The team had a total of two and a half hours to take the vehicle from a staging area to the launch pad, complete both flights, and return the vehicle to the staging area. The Level Two contest was similar, except the vehicle had to stay in the air longer (180 seconds), and both landing pads contained craters and rocks to simulate the lunar surface.

The 2006 X Prize Cup event included demonstration launches of privately built rockets.

Ambitiously, Armadillo planned to try winning the $350,000 Level One prize on Friday and the $1 million Level Two prize on Saturday. A vehicle called Pixel was up first, and its twin, Texel, would compete at the next level.

On Friday, October 20, the Armadillo team took Pixel to the launch pad. They ignited its engine. Using the familiar joystick controller, Carmack guided the vehicle gently into the air. He stopped the ascent when it reached a height of 180 feet, and he began the horizontal maneuver. As Pixel arrived over the landing pad, Carmack brought it down. This was the farthest and highest Pixel had ever traveled. The descent was faster than planned, however, and the vehicle hit the concrete pad hard. All four of Pixel's legs broke off.

"This was a successful qualifying leg for the prize, and I was preparing to fly it back 'on bloody stumps,' but we also cooked the drive and feedback cables . . . so we couldn't make the return flight," Carmack wrote in his online journal a few days later.

Working into the night, the crew repaired the damaged cables and added insulation to protect them from the heat. They recalibrated the control system to slow the vehicle's descent rate. And they took Texel's legs off and put them on Pixel. By Saturday morning, Pixel was ready to fly again.

The first leg of the flight satisfied all the competition's criteria, and it descended gently enough. But Carmack struggled to center it over the landing pad. It came down at the edge of the pad, with two legs on the concrete and two in the sand. Pixel fell over.

This time, the vehicle was undamaged, so the Armadillo team took it back to the

Pixel kicked up some dust as it rose off the launch pad.

144

staging area and then to the launch pad for one more try. The flight went well and Pixel landed completely on the pad, but the impact broke off one of the legs. Carmack decided to try to complete the return flight. The crew was not allowed to reattach the leg, so they simply used it to prop the vehicle up to a level position. When the engine ignited, though, the leg fell away and the vehicle tipped over, triggering the abort system and shutting down the engine.

"Sigh," wrote Carmack.

2007 X Prize Cup

The next annual X Prize Cup competition took place October 27 and 28, 2007. This time, the event was held at Holloman Air Force Base. The location change made it possible to provide the spectators with military aircraft flyovers and demonstrations between the relatively infrequent Lunar Lander Challenge launches. Total attendance for the two-day event was 80,000.

The second reason for the location change was financial. "The facilities they have here in terms of security, in terms of construction, are first class," Diamandis told a *Popular Science* reporter. "Instead of spending our money building landing pads and porta potties, we can spend that money bringing in more hardware and giving away more prizes."

Once again, Armadillo Aerospace was the only team competing, and the team members were ready to go. They brought Pixel again, but they were going to use Mod, a differently designed vehicle, for the competition. On the first morning, they set Mod on the launch pad but could not get the engine to start. They went back to

Mod's last attempt was colorful.

145

Veteran competitor Mod performed like a champ.

146

the staging area and cleared out a clogged igniter. On the afternoon attempt, they had clean ignition, and a perfect flight and landing. When they examined the vehicle on the landing pad, however, they discovered that the igniter was clogged again. It took them half an hour to figure out how to fix it on location; finally, they manually filed down a paperclip so it would fit into the igniter and clear the blockage.

Now they were ready for the return flight. The engine started with a bang and the vehicle hopped slightly, but the engine kept firing. Carmack really wanted to complete the round trip, so he decided to go ahead with liftoff. The exhaust plume did not look quite right, but the vehicle followed his commands and flew back over the original launch pad. He brought it down near the ground, trying to hover long enough to meet the 90-second limit. Small pieces had been flying off the vehicle during the return trip, and now another chunk blew off, rocking the vehicle. It tilted far enough to trigger the abort system, and Mod fell to the ground 8 seconds too soon.

Overnight, the team replaced some parts and installed a filter to keep the igniter from clogging. The next morning, they brought Mod back to the launch pad. The engine started with a loud bang, and as the vehicle rose into the air, unburned fuel was pouring out of the engine. Carmack brought the ship back down to the launch pad. Back at the staging area, they cleaned components and installed a new engine.

The afternoon offered one last chance. But when Mod's engine was ignited, it exploded and pieces flew in all directions. Carmack reported finding a large chunk of graphite from the engine stuck in the ground more than 200 feet from the launch pad. "We failed," he wrote two days later. "I haven't read any of the event coverage yet, or even gone through my rocket related email yet, because I am still sulking a bit."

2008 Lunar Lander Challenge

Funding and scheduling problems resulted in no X Prize Cup event being held in 2008, but the X Prize Foundation conducted the Lunar Lander Challenge competition October 24 and 25 at the Las Cruces International Airport. With no other events or demonstrations scheduled, the public was not invited. Although the spectators would be absent, the competition field was larger. For the first time, Armadillo Aerospace was not the only team flying a vehicle.

TrueZer0 brought its vehicle, named Ignignokt, from Chicago to try for the Level One prize. The TruZer0 team members were new kids on the block, having spent only ten months building their ship. It had successfully performed tethered test flights but had not yet tried a free flight or the maneuvering that the Challenge required. Still, the team gamely gave it a try. Ignignokt's liftoff was promising, with only a small amount of wobbling. As it rose past the qualifying height of 165 feet, however, it began spinning and wobbling more, finally tipping over. The ground crew shut off the engine, and the craft crashed to the ground, starting a small brush fire that was quickly extinguished. The flight lasted 19 seconds.

Next up: Armadillo. Carmack guided Mod up from the launch pad and over to the landing pad, and began to slowly lower it. He wanted to stop at a height of about

30 feet and hover long enough to meet the 90-second requirement, but Mod did not stop. It touched the landing pad before Carmack could reverse its descent, so the flight did not last long enough to satisfy the rules.

Armadillo had time for another attempt, and the first leg of the flight went perfectly. Mod settled down within 2 feet of the center of the landing pad, after being in the air more than 90 seconds. Because of airport activity, the team had to delay their return flight attempt until later in the day. When they did, the flight went perfectly. After years of perseverance and occasional frustration, the Armadillo Aerospace team won the Level One first prize.

Carmack and his team wanted to celebrate, but they also planned to fly Pixel the next day in an attempt to win the Level Two prize. They decided to just go back to their hotel. "After we had an opportunity to shower off the gritty accumulation of sunscreen, champagne, and New Mexico desert dust, everyone from the event showed up at the hotel, and they had a stack of pizzas delivered," Carmack wrote. "Perfect."

Pixel's attempt the following morning fizzled. "The ignition went fine, and it came up to idle," Carmack said in an interview after the attempt. "Then when we throttled up to full throttle, the LOX [liquid oxygen] valve moved fast and the fuel valve moved slow, so it went really lean and burned through the [engine]." The craft fell over off the launch pad.

In 2009, the Lunar Lander Challenge competition was modified so the teams could make their attempts at sites of their choosing. In September 2009, Armadillo Aerospace's vehicle, Scorpius, qualified for the Level Two prize in Texas. Two weeks later, Masten Space Systems' craft, Xombie, won second place in the Level One challenge in California, earning $150,000. Three weeks after that, another Masten's vehicle, Xoie, qualified for the Level Two prize. Although Armadillo had qualified first, Masten's landing accuracy was better. So Masten won the Level Two first prize of $1 million, and Armadillo came in second, winning $500,000.

Delayed, Not Dead

In 2005, when the Countdown to the X Prize Cup brought national attention to New Mexico, officials expected to start construction on the state's spaceport in 2007. That did not happen. In fact, by late 2008, when NASA held its last Lunar Lander Challenge competition in New Mexico, construction had still not begun. The vision had not faded, but the planning and approval processes proved to be slower than expected.

The annual X Prize Cup and Lunar Lander Challenge events helped keep the spaceport idea alive and exciting. In 2007, Homans said in an X Prize Foundation press release, "The Cup remains the only annual event where the entire family can see the next generation of private spaceships in action, and that's an important distinction for the state, Las Cruces and Doña Ana County as the host community."

Resources

Belfiore, Michael. *Rocketeers: How a Visionary Band of Business Leaders, Engineers, and Pilots Is Boldly Privatizing Space.* New York: HarperCollins, 2007.

Carmack's reports on X Prize Cup competitions. Accessed at http://www.armadilloaerospace.com/n.x/Armadillo/Home/News.

Diamandis interview on *The Space Show*, August 8, 2001, rebroadcast August 24, 2007. Accessed at http://thespaceshow.com/detail.asp?q=759.

Diamandis interview on *The Space Show*, August 8, 2005. Accessed at http://thespaceshow.com/detail.asp?q=372.

Homans interview on *The Space Show*, May 11, 2007. Accessed at http://thespaceshow.com/detail.asp?q=712.

Chapter Nine: Spaceport America

Commercial spaceports supporting privately owned and operated spacecraft offer a variety of opportunities. For example, NASA planned to retire its Space Shuttle fleet in 2011. After that, they have to ferry astronauts and cargo between Earth and the International Space Station aboard a Russian vehicle or on a commercial vehicle operating from a US spaceport. Since at least the mid-1990s, NASA has encouraged private development of space operations. The 1998 final report of a study conducted by NASA and the Space Transportation Association (STA), a commercial trade group, concluded that "The opening of the frontier of space—not just to government missions and astronauts, but now to private individuals and private sector businesses—is a space challenge of overarching importance. It is especially important for the democratic United States of America."

Cost was a major factor. According to the report, "The cost of a Shuttle trip to/from orbit for a half-dozen people (and upwards of 40,000 pounds of cargo) now approximates $400 million, the individual Shuttle trip turnaround time is about a half-year, and the possibility of a fatal accident about 1% ." The study group expected that booking passage on a commercial flight would reduce the cost by a factor of ten in the first generation of technological development and another factor of ten in the second generation.

The study group also expected far more frequent flights to be possible, and the safety factor to be "several factors of ten" better than the government's Shuttles. It may seem a stretch to expect commercial spaceflight to be far safer than NASA's Space Shuttles, given the mishaps that plagued the Lunar Lander Challenge entrants. On the other hand, developers new to spacecraft design made tremendous strides between 2006 and 2009. Those talented and dedicated entrepreneurs continue to develop their innovative vehicles at a rapid pace, largely unfettered by government bureaucracy and always keeping safety paramount.

While NASA hoped to take advantage of the commercial development, the report focused on space tourism as a driving force in that development. "Since any US next generation (post-Shuttle) space transportation vehicle fleet must be privately funded, it must be designed to serve commercial as well as government markets. The commercial markets must be large enough to make investment profitable; space tourism is potentially the earliest and largest of these markets," the report stated.

That vision of complementary roles involving scientific missions and space tourism was at the heart of New Mexico's plan to build a spaceport. "We want to become the launch pad for this new industry, not only for launches and operations, but for research and development, testing, manufacturing, maintenance and all of the tourism impact as a result of having this spaceport in New Mexico," Rick Homans told the *Business Xpansion Journal* in May 2006. Nearly a decade after the NASA/STA study, tourism still seemed to be a major funding source for commercial space development.

Market Analyses

During the early 2000s, three major studies evaluated the viability of space tourism as a commercial enterprise. In 2002, Futron Decision Management Solutions, a consulting firm focused on the aerospace industry, hired the Zogby International polling firm to conduct such a study. Reasoning that, at least initially, space tourism would be quite expensive, Zogby pollsters interviewed 450 individuals who had an annual income of at least $250,000 or a net worth of at least $1 million. The study discovered viable markets for both suborbital and orbital recreational flights. In particular, assuming a ticket cost of $100,000, they projected that suborbital flights would attract 503 passengers during the first year of operation. Further assuming that the cost would remain constant during the first five years of operation and incrementally decrease to $50,000 over the following decade, they projected that 15,712 people would take recreational suborbital flights during the fifteenth year of operation. Those numbers translated to annual revenue starting at $36 million and reaching $786 million during the fifteenth year of operation.

The pollsters found that the market for orbital tourism flights was considerably smaller. They envisioned a two-week flight preceded by six months of training. Assuming an initial ticket price of $20 million, gradually declining to $5 million, they projected that a total of 419 people would participate during the first fifteen years of operation.

In addition to the financial predictions, the Zogby researchers asked the poll participants about the importance of certain aspects of a suborbital flight. Two-thirds of the people said that seeing the Earth from space was "very important" to them. About 40 percent thought it was very important to experience weightlessness or to "experience what only astronauts and cosmonauts have experienced." Only one-third said experiencing the acceleration of a rocket launch was very important.

The interviewers asked the survey participants whether they thought several activities were "high risk," "medium risk," or "low risk." Less than half (45 percent) thought space travel was highly risky. In contrast, 57 percent considered mountain climbing a high-risk activity, and 72 percent said skydiving was highly risky.

In 2005, Economic Development Department Secretary Homans also became the chairman of the state's Spaceport Authority. He asked both Futron and New Mexico State University to study a spaceport's economic impact on southern New Mexico. Both concluded that a spaceport would generate thousands of jobs and hundreds of millions of dollars of revenue over the next five to fifteen years. The jobs and revenue would come from not only spaceport operations, but also new support businesses such as rocket component manufacturers, spillover effects in the general economic activity in surrounding communities, and increased tourism.

In addition to the economic impact of a spaceport, the NMSU study looked at political, social, and technological factors. "Perhaps the most significant social factor associated with the commercial space industry is its educational value," it concluded, noting that the education benefits could be nearly as important as the economic impacts. "Interest among students in science, technology, engineering, and mathematics (STEM) was at its highest during the Apollo-era and has languished since. This endangers the nation's ability to go back to space, both militarily and commercially. . . . While the spaceport by itself cannot correct all of these problems, it can be a key to turning it around."

Spaceport America Realized

When the New Mexico legislature met in January 2006, space tourism was no longer the theoretical notion examined in the 2002 Zogby poll. The more recent studies, coupled with SpaceShipOne's success in the X Prize competition, made commercial space ventures appear to be just over the horizon.

The 2006 legislature appropriated $110 million toward building the Southwest Spaceport, soon to be renamed Spaceport America. Over the next two years, local road construction funds and voter approval of small increases in the gross receipts taxes (essentially sales taxes) of Doña Ana and Sierra counties cobbled together the rest of the $200 million needed to build the spaceport. Because education was one of the three primary objectives of the program (along with economic development and tourism), one-fourth of the income generated by the additional gross receipts taxes was earmarked for supporting math and science education in those two counties. This would encourage and prepare students in communities near the spaceport for careers in the aerospace industry.

Spaceport America is not the only commercial spaceport licensed by the Federal Aviation Administration (FAA), nor was it the first. When it obtained its license in December 2008, New Mexico's spaceport joined licensees in Alaska, California,

"A 4,700-foot elevation . . . drops the cost of reaching earth orbit by up to $90 million, compared with launching at sea level."—*Discover* magazine, in listing the groundbreaking for Spaceport America one of the "top 100 events poised to change the world in 2010."

Oklahoma, Texas, and Virginia. A commercial spaceport in Florida received its FAA license in early 2010. However, Spaceport America is the only one to be purpose built for commercial use; all other US spaceports use current or former military facilities.

Location, Location, Location

For most of the same reasons the US government established the White Sands Missile Range in the Tularosa Basin, the state of New Mexico placed Spaceport America a few miles west of WSMR. Normal weather patterns will allow launches 340 days a year. The air is dry and noncorrosive. The elevation gives launches a fuel-saving "leg up." WSMR's restricted airspace to unlimited heights keeps airplane activity safely away from the launch sites. The sparse population around the spaceport site results in reduced risks and lower insurance rates, and the Spaceport Authority intends to keep it that way.

"We are actually looking at working with Sierra County (where the spaceport is located), the state land office, the Bureau of Land Management, and private landowners to create a 20-mile radius around the spaceport where any commercial development would be restricted or prohibited or limited in whatever way possible," Homans told a 2007 radio audience. "We would like to see the commercial development—the hotels, the manufacturing, the R&D—go in neighboring communities of Truth or Consequences, Las Cruces, Hatch, Alamogordo, and the spaceport really be a true operational spaceport without a lot of ancillary development built up immediately around it."

Another important advantage of being near WSMR was all of the rocket tracking and recovery infrastructure and experience the operators of that facility had. Homans worked hard to build a true partnership with WSMR rather than a subservient, customer-type relationship. "White Sands has an interest in Spaceport America because, quite frankly, their government business is on the decline, and they need to be looking for new sources of revenue to maintain their operations there on this huge and historic range," he said. "So with a commercial spaceport going in right next door, they see the opportunity for new customers in the decades to come. We see the opportunity to not have to invest in all kinds of infrastructure that's in place 20 miles away."

Homans explained that WSMR tracks the spaceport's unmanned rocket launches. "They provide the restricted-access airspace that's so important, and they also are working with us to provide range recovery services if the vehicle comes down on the range itself," he said.

A Home on the Range

The Spaceport America site is an irregularly shaped, 18,000-acre parcel (28 square miles). The state owns most of the land and leases it to two ranching families. Ben and Jane Cain operated the Bar Cross Ranch, and their daughter and son-in-law, Judy and Phil Wallin, operated the Lewis Cain Ranch (Ben Cain died in 2007, and the Wallins' daughter Amanda eventually took over managing both ranches). Each

154

family owned a small portion of their land but leased most of the grazing land from the state and federal governments.

The Cains' first home, the Buckhorn Ranch, was on the east edge of the San Andres Mountains. They were forced to move during World War II when the federal government prepared to test the first atomic bomb in that area. They re-established their business on the west side of the San Andres range, and eventually operations on the White Sands Missile Range encroached on their ranch land. Once or twice a month, the Cains were paid to leave their homes for six to twelve hours during a WSMR test. "They never really do tell you why," Jane told the author of *Tales from the Journey of the Dead*. "But we go to town and make it our monthly shopping trip."

Homans was determined to treat the two families fairly and respectfully. "They had come out of being, in their minds, mistreated by the federal government, when it came to the Trinity Site and their ranch lands over there and being kicked off and moved away," he said in a 2010 interview. "And then here, fifty years later, was some other government program coming into their environment and on what they considered their property, even though it was leases on state and federal land." After having the ranches appraised, Homans offered to purchase them for the full appraisal value of more than $4 million, but the families wanted to stay in business on their long-time homelands. He respected their decision and took a different approach, asking if it might be possible for them to coexist.

To a person looking toward the east, Spaceport America's terminal building looks like a gentle rise in the desert landscape.

They were able to work out an agreement by which the ranchers kept their grazing leases while the Spaceport Authority obtained a business lease on the part it needs for the spaceport. The spaceport's buildings and runway occupy a small portion of the joint-use area, and the ranchers can continue to graze cattle on most of the land. In addition to making lease payments to the state, the Spaceport Authority pays the ranchers a yearly fee for inconveniences caused by spaceport operations, which may include evacuating areas around vertical launches. The spaceport also bears the cost of relocating, when necessary, any structures such as cattle pens and watering facilities.

When the Cain and Wallin families signed the spaceport lease agreement in late 2006, they had lived and worked on that land for five decades. That short period pales in comparison to the age of another nearby neighbor. El Camino Real de Tierra Adentro (the Royal Road to the Interior), though not a living resident, is a valuable fixture that crosses the spaceport land, near enough for the terminal building to be visible from the trail, which was established in 1598. In 2007, the National Trust for Historic Preservation listed El Camino Real among the eleven most endangered historic places in the United States because of the spaceport's construction. Calling the trail the earliest Euro-American trade route in the country, the description on the organization's website said, "Camino Real is considered by many to be one of the largest and most important artifacts of the Spanish Colonial era in the United States and one of the most valuable single markers of the Hispanic experience in the Southwest."

Not only might the spaceport's terminal facility mar the historic view toward the east from El Camino Real, but the access road from Truth or Consequences to the spaceport crosses the historic trail. Preservationists feared the increased traffic and its attendant air pollution would significantly affect the landmark. The Spaceport Authority took seriously its responsibilities to respect this important state relic. In a June 2007 press release, the Authority summarized its accommodations: "These include: moving the Spaceport runway almost three miles from the Trail, not making any new crossings of the Trail, attempting to limit development in a twenty-mile radius of the Spaceport, designing the site and buildings to blend into the desert landscape, burying utility lines, and bringing visitors to the site in hydrogen-powered buses, not single-occupancy vehicles."

Highlighting the juxtaposition of the futuristic spaceport with its four-century-old transportation ancestor, the spaceport's groundbreaking ceremony in June 2009 included costumed representatives of the Spanish colonizers. They marched from the direction of the trail and presented New Mexico Governor Bill Richardson with a scroll that proclaimed, "The explorers of the past now extend our best wishes to the future explorers of the trail to space."

UP Aerospace

Spaceport America became operational even before its groundbreaking ceremony. UP Aerospace, the spaceport's first tenant to launch from the site, builds rockets and launches them vertically. The facilities the company leases at the spaceport are much simpler than the horizontal launch facility and its associated terminal hanger building. Located 5 miles away, the vertical launch site consists of three small buildings and a 25-foot by 100-foot concrete pad with a 56-foot tall tower that can be rotated and tilted to desired launch directions.

Jerry Larson, a former Lockheed Martin engineer, started Denver-based UP (pronounced *up*) Aerospace to develop and build rockets capable of carrying commercial payloads inexpensively and frequently. He told a *Denver Post* reporter in 2010 that his three-person crew can launch a rocket for less than one-tenth what a NASA launch would cost, partly because a NASA launch requires a crew of about fifty people and partly because Larson builds his rockets himself. He also cited Spaceport America's low-cost vertical launch facility and the fact that there is less "red tape" than at a federal facility. Those factors decrease the costs and the amount of time it takes to arrange a launch. Arranging and accomplishing a launch takes only three months, one-sixth the time it would take at Vandenberg Air Force Base. "We can do it faster and cheaper while ensuring that safety is paramount," Larson told the reporter. "That's what Spaceport America is all about."

Larson launched his first 20-foot-long, 10-inch-diameter rocket from Spaceport America in September 2006. It carried several dozen payloads for a variety of clients. Students, ranging from elementary school to college and from across the country, had devised experiments such as comparing the effects of suborbital flight on analog and digital watches. The widow of a Colorado veterinarian packed in a small portion of her late husband's cremains. After a technical delay of several hours, the rocket was ready to go. Rick Homans hit the launch button. The rocket zoomed straight into the sky for 9 thrilling seconds. Then its path became erratic, and the rocket reached a height of only 42,000 feet—just one-eighth of what was planned. The rocket plummeted to the ground, crashing at 700 miles an hour. The experiments were destroyed. The capsule containing the veterinarian's ashes broke open, spilling them on the ground.

At least some of the observers chose to accentuate the positive aspects of the failed flight. "One of the unique parts about what we're doing here in New Mexico is, pretty much everything that we're doing is transparent and open," Homans said. "We didn't hold a bunch of stealth launches to try to get all the systems down. . . . There is some trial and error, there are some lessons that we're learning as we go forward, and I think we'll get better with each launch that we undertake."

The veterinarian's widow said, "He loved the Earth and the sky and all the creatures in between, and I think the fact his ashes were scattered into the Earth is a fitting conclusion to his life."

The Wallins' ranch house is only a mile from the vertical launch pad. "What's it like to see it go up?" Judy Wallin said to a reporter in 2009. "It's 'chick-koom,' and it's gone. It is exciting."

UP (Higher) Aerospace

Despite the rocket's failure, the possibility of economical and accessible suborbital flights impressed potential customers. Payload space on UP Aerospace's next two flights sold out. Larson analyzed the problematic first flight and added a fourth tail fin to the rocket to stabilize its flight. The second launch, in April 2007, carried forty-five scientific experiments prepared by hundreds of students. For instance, a group of students from Hatch, New Mexico, packaged some Big Jim chile seeds for the flight and planned to later compare those seeds' germination with control seeds that did not travel into space. In all, the rocket could carry 110 pounds of payload, and several commercial ventures were on board, too—"companies literally launching new products that they'll unveil after the launch," said UP Aerospace CEO Eric Knight. One company's service was to send personal photographs into space and return them to their owners as mementos.

Another company, Celestis, arranged space flights for the dearly departed. Among the 200 "passengers" on this flight were small portions of the cremains of James Doohan, the actor who portrayed Scotty on the popular *Star Trek* television series, and Gordon Cooper, an astronaut who orbited the Earth twenty-two times in 1963 during NASA's final Mercury Program flight. Celestis' CEO, Charlie Chafer, was pleased for his company's cargo to share the payload capsule with students' experiments. "The key to the future is the kids, and there's nothing like 'hands on,' there's nothing like really saying, 'that's my rocket out there, I've got something on that' to keep them motivated," he said on launch day.

Together, the widows of Doohan and Cooper pressed the launch button. The rocket soared skyward, peaking at 73 miles above the Earth. The payload capsule landed, intact, on the White Sands Missile Range. "Spaceport America is now a real spaceport," Larson said.

For two weeks, weather problems prevented the recovery crew from reaching the payload capsule, which had landed in rugged terrain accessible only by helicopter. After the capsule was recovered, Homans said, "Now we can all say 'Mission Accomplished.' Scotty and Gordon Cooper's final trip to space took a little longer than we expected, but we can be sure they had a beautiful ride and we welcome them and all the other astronauts back to Planet Earth and to New Mexico."

Since that first successful flight, UP Aerospace continues to launch payloads for a variety of commercial enterprises, including Lockheed Martin. In October 2009, UP launched an unmanned, one-fifth scale model of a reusable rocket plane. Slater Voorhees, lead project engineer for Lockheed Martin Space System's Advanced Programs, told *SPACE.com* that studies and theoretical analysis can go only so far when developing new technology. "We wanted to take it one step further . . . to get our engineers using hands-on hardware," he said. "There's just so much you can do when you simulate things, study things, and do PowerPoint presentations."

"There is a growing interest in more launches coming from the US Air Force, Lockheed Martin and others for next year," UP Aerospace's president Larson said after the rocket plane test. In fact, in May 2010 UP conducted a suborbital flight for the Department of Defense's Operationally Responsive Space (ORS) Office. The payload was an instrument package built by another private company, the Schafer Corporation, in only a week. After the launch, the ORS Office's director, Peter Wegner, said, "This launch proved to be a very cost-effective way to demonstrate key ORS enabling models of rapid development and build of a payload, integration and test of the payload, and identification and assurance of payload technical readiness."

Payload Specialties

Payload Specialties, another Spaceport America tenant, offers services similar to those of UP Aerospace. It invites individuals to send personal items on its launches. "Offering payload space to the general public is quite unique," says its website. "Have an item of interest poised in a display cabinet, a model rocket, poster, figurine, business card, tie, etc. Have it launched into space. Make your items even more of a conversation piece."

One of Payload Specialties' commercial customers is Heavenly Journeys. Like Celestis, it prepares cremains for space flights, and it offers an additional option: space burial. Using a specially designed space flight urn and dispersal system, the company is able to scatter an individual's ashes "across the heavens." A description on the company's website says, "Because Heavenly Journeys flights go into suborbital

The east side of the terminal offers sweeping views of the runway and the scenery.

space, your loved one's cremated remains will eventually return to Earth. However, due to the extreme elevation of dispersal, the cremated remains may stay aloft for weeks or even months."

Microgravity Enterprises, a New Mexico company and tenant of Spaceport America, sends ingredients into space aboard Payload Specialties rockets and then offers them for use in a variety of products including vitamin supplements, ice cream, cosmetics, sports equipment, and solar cells. One product, Antimatter Energy Drink, is manufactured and sold by Microgravity itself. The company's website touts its praises: "This stuff has been ripped through the ozone; it just doesn't get any more extreme than that. . . . This is not just some energy drink that will give you a little boost. It's Rx for the Cortex."

Armadillo Aerospace

After winning prizes in both levels of NASA's Lunar Lander Challenge, Armadillo Aerospace continued to develop its line of reusable, vertical-launch rockets. In April 2010, Space Adventures Ltd, a company that arranges adventure trips like private passengers going to the International Space Station, signed an agreement to book seats on suborbital flights on vehicles Armadillo was still developing. "A decade of research and development has gotten us to the point where we can credibly talk about commercial passenger experiences," Armadillo's president, John Carmack said. "Everything is coming together—there is enough clarity in the technical, regulatory, and market factors that it is the right time to form a solid partnership with Space Adventures to help us take things through to commercial operation."

An Anchor Tenant

The vertical-launch Armadillo flights offer an alternative experience to the horizontal-launch commercial flights offered by Spaceport America's anchor tenant, Virgin Galactic. After Burt Rutan won the X Prize, Sir Richard Branson, the British owner of Virgin Galactic, contracted with him and his backer, Paul Allen, to develop a second generation of WhiteKnightOne and SpaceShipOne for space tourism flights and sell them to Virgin Galactic. Branson announced the new venture on October 4, 2004, right after the flight that won the X Prize.

"Just like with the X Prize Cup, we said that it was absolutely essential that we recruit Branson to operate out of New Mexico," Homans said in 2010. After some preliminary discussions between representatives of New Mexico and Branson's company at space conferences, Homans and an associate flew to London to meet with Alex Tai, Virgin Galactic's vice president for operations. "They told me later that they had initially agreed to the appointment just out of courtesy, because they just didn't want to be rude," Homans said. "Then during the meeting, Alex became convinced that this thing had a lot of merit, and so he brought in Will [Whitehorn, Virgin Galactic's president]." The meeting lasted several hours, and serious negotiations continued for more than a year.

In December 2005, Branson and Governor Richardson signed an agreement.

Virgin Galactic would lease 84 percent of the 100,000-square-foot spaceport terminal building; moreover, it would move its international headquarters to the site. "This flat, deserted, mostly rain-free stretch of New Mexico is the perfect location," Whitehorn told a *New York Times* reporter. "It is about the closest you get on planet Earth to a Martian landscape."

Virgin Galactic's original lease agreement extends for twenty years, starting when the terminal building is complete. Besides the Virgin Galactic corporate offices, the terminal building also houses offices for the New Mexico Spaceport Authority and other tenants, hanger space for the Virgin Galactic fleet of two motherships and five spaceships, training and lodging spaces for its customers, mission control, and visitor observation areas. Construction on the terminal building began in the spring of 2010.

The terminal is partly buried and increases in height from its ground-level western roof to its east side, where the undulating ceiling is as high as 50 feet. Floor-to-ceiling windows cover the east wall, and skylights in the roof bring natural light into the building.

What Is Virgin Galactic?

Virgin Galactic is one of more than 300 companies within the Virgin Group created by Richard Branson. Branson and a few friends started the first of these companies in 1970, when he was seventeen. They decided to set up a mail-order business to sell record albums at discounted prices. As they were tossing around ideas for what to name the company, one of the group said, "Virgin, because we're complete virgins at business." The year after starting Virgin Mail Order, Branson opened a Virgin Record Shop in London, and soon afterward he opened a recording studio and started the Virgin record label.

Twin-cabin WhiteKnightTwo carries SpaceShipTwo (center) high into the sky before releasing it—just as Rutan's earlier version did to win the X Prize.

161

In 1984, Branson's companies began heading skyward. He started his first airline company, Virgin Atlantic. The following year, he began a decades-long series of personal adventures, including record-breaking attempts in various types of transportation—sail and powered boats, hot-air and gas balloons, and even an amphibious vehicle. During one of those record tries, he began to envision Virgin Galactic and space travel. In 1995, he was waiting for the right weather conditions for what turned out to be a record-setting, nonstop balloon flight from Morocco to Hawaii. He got into a conversation with Buzz Aldrin, who had been the second human to set foot on the moon when he landed there with fellow NASA astronaut Neil Armstrong in 1969.

In a 2009 *AstrocastTV.com* interview, Virgin Galactic CEO Stephen Attenborough described the conversation: "Richard started talking to Buzz, saying, 'It's interesting and a little frustrating because when I was a kid in the '60s, my mum told me that everybody would be going on holiday into space by 1982 because the progress at that point in time was just enormous. It hasn't happened, and why not? And can we do something about it?'" They discussed some ideas about how to make commercial space travel viable, including the potential for launching a ship from an airborne carrier vehicle to increase safety and reduce costs. Branson became so intrigued with the notion that he told Virgin Group staff members to start building a file about new space launch technologies, and in 1999 he registered the company name Virgin Galactic with British authorities.

The X Prize competition had already been announced, but progress on new vehicles was slow. Things changed in 2003 when Branson learned that Burt Rutan was building a spaceship. He knew that the unique composite material Rutan used to build his vehicles was very light and very strong. And he knew of Rutan's reputation for designing innovative, very high-performance aircraft. He negotiated with Rutan and his backer, Paul Allen, to become a sponsor of SpaceShipOne. Hurriedly revving up the Virgin Galactic company and getting a logo designed, Branson managed to get the company's symbol on SpaceShipOne in time for the two X Prize-winning flights in 2004. In the following months, he struck deals with Allen to license the ship's design and with Rutan to design and build larger versions, WhiteKnightTwo and SpaceShipTwo, that could carry half a dozen tourists on a suborbital space flight.

"Of all the projects that I've worked on in my life, of all the businesses I've started, this is by far the most exciting," Branson said in Santa Fe, a few weeks before the New Mexico legislature convened in January 2006. "This business will mark a milestone in world history, and it will launch a new space industry—a private space industry, driven by innovators and entrepreneurs and new technologies and bold thinkers."

Gemini and Apollo astronaut Buzz Aldrin said in a 2010 interview that he thought the term *astronaut* should be reserved for those who travel into space during their military service. He suggested *star traveler* as an appropriate term for private spaceflight participants.

162

Who Rides SpaceShipTwo?

People who fly on Virgin Galactic trips are not called *passengers*. They are *participants*. The US Congress, in the Commercial Space Launch Amendments Act of 2004, defined the term by stating "'space flight participant' means an individual, who is not crew, carried within a launch vehicle or reentry vehicle." As a practical matter, the difference implied by this distinction is that the participant understands and accepts the inherent risks of space flight. The Commercial Space Launch Amendments Act required that space flight participants be fully informed of the flight's risks as well as requirements such as physical exams and flight training, and that they sign an informed consent form before participating.

By the fall of 2010, more than 380 people had reserved seats on upcoming Virgin Galactic suborbital flights, by either paying the full ticket price of $200,000 or putting down a deposit ranging from $20,000 to $175,000 (full price or larger deposits earned an earlier flight date). Some are celebrities, such as actress and entrepreneur Victoria Principal. At a 2005 press conference in Santa Fe, Principal said, "For me, this is a dream come true, to go into space and look back and see the Earth, my home."

Other participants are not public figures. Chemist Lina Borozdina-Birch works for a pharmaceutical company. She and her husband took out a second mortgage on their San Diego home to buy her ticket. "I'm so excited," she said at the spaceport's 2009 groundbreaking ceremony. "Up until this point in my life, [going into space has] been like a science fiction novel. It's worth me paying my mortgage for the next 20 years twice to have my dream come true." In a 2007 interview with a *USA Today* reporter, she admitted to having a fear of flying. "I'm going for the view. I'm not going for the thrills," she said. "I'm working on not freaking out on the flight."

Branson has offered free tickets to two famous people who take the opportunity seriously. One of them, scientist and avid environmentalist James Lovelock, told a reporter in 2007 that he expected space tourism to help protect the Earth. "The view that all these people will get will show them what an incredibly beautiful planet we have," he said. "And they'll take back the story, and that story will tell people, the rest of us on Earth, just how important it is to look after our planet." The other free ticket recipient is astrophysicist Stephen Hawking. "I see great dangers for the human race. There have been a number of times in the past when its survival has been a question of touch and go," Hawking said in a 2010 interview. "The human race shouldn't have all its eggs in one basket, or on one planet."

Some participants won their tickets in ad-campaign contests. During half time of the 2005 Super Bowl, Volvo kicked off a promotion that invited people to register on line for a chance to win a free ticket on Virgin Galactic. Colorado resident Doug Ramsburg's name was drawn from the pool of 135,000 entrants. When he won, he said, "Growing up in Denver, surrounded by the Rocky Mountains, I would think, 'what's beyond these mountains?' Then, as I got older and would fly on airplanes, I would think, 'what's beyond this sky?' I never thought I'd actually get to find out. Now I will."

A Virgin Galactic ticket was offered as grand prize in several other contests or promotions, including a 2008 Virgin America airline drawing for qualifying frequent flyers, a 2008 campaign launching Virgin Atlantic Airway's branded American Express credit card, and a 2009 Guinness Beer contest celebrating the brewery's 250th anniversary. In 2006, British businessman Alan Watts bought a Virgin Galactic ticket by redeeming two million frequent flyer miles he had accumulated with Virgin Atlantic. "It's going to offer the best view in the world, and I can't wait to make history," Watts said in a press release. "The nearest I've come to space before was going on the Space Mountain ride in Florida."

Preparing for the Flight

Participants first have to get medical clearance from their personal physician. When they arrive at Spaceport America, they will experience three days of flight preparation, including further medical evaluation. Branson is confident that the vast majority of candidates will meet the physical requirements. In 2007, eighty-one of Virgin Galactic's 100 Founder customers (who paid the $200,000 in full) underwent preliminary training on a centrifuge that simulated the g-forces they would experience during the flight. The two days of training included classroom preparations and six rides on the centrifuge, duplicating the 3.5 g's during SpaceShipTwo's launch and 6 g's during reentry. The Founders who took part in this preliminary training exercise ranged in age from twenty-two to eighty-eight, and 93 percent of them successfully completed the training.

When they arrive at Spaceport America, participants will begin what Virgin Galactic's website describes as "three days of pre-flight preparation, bonding and training." (Other activities will be available for family members who come with them.) The participants will learn techniques for staying as comfortable as possible while experiencing the g-forces of acceleration and reentry. They will also get instructions for enjoying weightlessness. Part of the training will include riding in WhiteKnight-Two (WK2), the airplane-like vehicle that carries SpaceShipTwo (SS2) for the first portion of its flight. The interior of one of WK2's twin cabins is just like the interior of SS2, and the carrier vehicle can duplicate the g-forces a ride in the spaceship generates. "Furthermore, with its powerful spoiler flaps, WhiteKnightTwo can also duplicate SpaceShip's approach flight path angle, making it a highly useful in-flight simulator for this important part of SpaceShip's mission," according to the Virgin Galactic website.

Riding into Space

In 2007, Virgin Books published *Destination Space: How Space Tourism Is Making Science Fiction a Reality*, by Kenny Kemp. Richard Branson wrote the foreword for the book, which tells the story of Virgin Galactic and its future passenger flights. About the same time, animated videos depicting passengers' experiences on Space-ShipTwo flights began appearing on websites like *YouTube*. The following description is based on those predictions:

SpaceShipTwo carries six space tourists and two pilots. As the trip begins, the ship is attached to WhiteKnightTwo. The combined vehicles zoom down Spaceport America's 10,000-foot-long runway and soar into the air. For the next hour, WK2 carries SS2 upward in an increasingly wide spiral to an altitude of 50,000 feet—10,000 feet higher than commercial airplanes fly. The pilot alerts the participants to prepare for launch and releases SS2 from the mothership. The spaceship falls freely for a few seconds, then the pilot fires the rocket engine and pulls the ship's nose up. For about a minute, the force of acceleration keeps everyone pressed against their seats. By the time the pilot shuts the engine off, the craft is traveling 2,500 miles an hour—three times the speed of sound.

For the next 6 or 7 minutes, the participants are free to move weightlessly around the cabin, surrounded by the absolute silence of space. They can float and flip effortlessly around the 7½-foot diameter cabin. They can look through one of nine large windows, down at the Earth 360,000 feet (68 miles) below, or out into the vast blackness of space. At the pilot's command, they return to their seats and fasten their seat belts.

The pilot raises the ship's rear wings up. This aerodynamic shape makes the ship fly like a badminton birdie, pushing its nose forward and down. The increasingly thick air pushes against the raised wings, slowing the ship's descent enough to avoid overheating the nose. When SS2 falls to 50,000 feet, the pilot rotates the rear wings back to their horizontal position and flies the unpowered ship back to the spaceport's runway, landing it like a streamlined Space Shuttle.

When the space adventurers step out of the ship and reenter the terminal build-

Folding SpaceShipTwo's tail section tips the craft's nose downward, pointing it back toward Earth.

ing, they slip away for a post-flight gathering. "You have just had the most incredible experience of your life," Julia Tizard, Virgin Galactic's operations manager, told the author of *Destination Space*. "You simply want to go away into a nice quiet room so that your body has time to reassess itself and calm down." The just-ended flight's pilots and participants use this time to share their thoughts and emotions, relax, and let their adrenaline levels return to normal. Then they rejoin their families and get ready to celebrate.

Environmental Aspects

In September 2006, Richard Branson and other Virgin Galactic officials unveiled the interior design concept for SpaceShipTwo's cabin in New York. At that event, Branson said, "If you're going to build a spaceship, you've got to build a green spaceship." That may seem like an inherent contradiction to people who worry that recreational spaceflight wastes resources and creates pollution. But several aspects of the Spaceport America and Virgin Galactic partnership are carefully designed to preserve resources and operate cleanly.

The terminal hanger building's earth-sheltered design reduces its heating and cooling demand by 60–70 percent by shielding it from the sun's rays. Fresh air enters the building through large tubes buried in the ground. Drawing the sun-heated air through the cooler ground reduces its temperature before it flows into the building during warm months, and pulling winter-chilled air through the warmer soil helps heat the building at other times. The building uses solar panels to generate electricity and is equipped with water recycling systems.

WhiteKnightTwo and SpaceShipTwo are both made of lightweight composite material, which reduces the amount of fuel necessary to fly them. WK2 uses commercial jet engines that are fueled by either kerosene or butanol, a biofuel that can be made from a variety of organic materials, including algae. Launching SS2 from high in the air offers several advantages. Its rocket engine burns for a much shorter time and uses much less fuel than a ground launch would require. The engine uses nitrous oxide (so-called "laughing gas") as the oxidizer and rubber as the fuel (future versions might use recycled nylon as fuel). Whitehorn told a *ScientificAmerican.com* reporter in 2009 that the per-passenger carbon dioxide emissions of a flight of WK2 and SS2 would be only 60 percent of the per-passenger emissions of a round-trip commercial flight between New York and London.

Furthermore, Branson believes the technologies being developed by commercial companies like Rutan's will have long-term benefits outside of space tourism. "Today's generation has the technological ability to do more industrial work up there, providing communications, advanced science and even, potentially, solar power and [computer] server farms in space—thus taking CO_2-intensive industry out of the atmosphere," he told a *Smithsonian* reporter in 2010. "The challenge is to get the technology up there in a safe, reliable, and cheap way with minimal environmental impact. Nonreusable rockets launched from the ground based on designs from the 1940s are not the answer."

Keeping it Real

Branson has a reputation for innovation and creativity in business and for taking risks in pursuing his personal passions for record-breaking adventures. But he is also a savvy business person who understands the importance of safety for the customers of his airline and railroad companies—and now of his space travel venture. Speaking at a February 2010 workshop in Colorado, Virgin Galactic CEO Stephen Attenborough said, "You cannot cut corners. Safety has to come first. The project has taken longer. It's more expensive . . . probably more complex than we thought it would be initially. But it's going very well."

A July 2007 event brought the issue of safety to the forefront. The Spaceport Authority held a competition to select an architect for the terminal hanger building, and Homans was traveling to Las Cruces with representatives of Virgin Galactic and the three finalist firms to announce the winner at a press conference. "We got to around Los Lunas and Alex Tai's cell phone went off, and that was the notification of the big explosion in Mojave, where three people were killed," he said in a 2010 interview. Three employees of Rutan's company, Scaled Composites, died and three others were injured in an explosion while they were testing the nitrous oxide flow system for SpaceShipTwo.

Subsequent investigations by California OSHA (Occupational Safety and Health Administration) and Scaled Composites were unable to pinpoint the cause of the explosion, but there were some indications that a contaminant might have been present in the nitrous oxide tank. Rutan's company made several changes in its employee training and nitrous oxide handling procedures to improve safety. "This was an industrial accident at Mojave spaceport, and it was not an accident of the new space industry or of the Virgin design," Homans said.

However, the terrible incident focused the attention of everyone involved with the spaceport project on the inherent risks of rocket launches. "We all know that we are working in a risky business and industry—launching people and payloads into space in ways that have never been done before," Homans said. "Someday, something's going to happen. . . . One of the big risks is, how do we survive a tragedy? How do we survive our own Columbia or Challenger disaster in this fledgling industry? What will the impacts be of that, to public support, to government regulation?"

All aspects of the design of WK2 and SS2 have been done with safety in mind. For example, the spaceship's air launch at 50,000 feet eliminates the need for large amounts of explosive fuels being used at ground level. The ship's solid rubber fuel and the nitrous oxide oxidizer are nontoxic and benign when handled properly. SS2's folding rear wing section automatically orients the craft for a safe re-entry, and no fuel is on board for the glide-in landing.

The 2007 explosion heightened the safety consciousness that the project's developers had from the beginning, and successive tests on WK2 and SS2 proceed only when they are ready. "There's a reasonable possibility that we could see the first drop flight in the fall, but as always, everything is predicated on thoroughness and safety," Attenborough told *Space.com* in July 2010. "No corners will be cut in order to achieve arbitrary deadlines." In fact, the first drop flight test took place on October 10, 2010. WK2 released SS2 at an altitude of 45,000 feet, and the spaceship glided for 15 minutes before landing safely on the runway at the Mojave air and space port.

Not Just Tourism

Virgin Galactic's dramatic space tourism flights captivate the public's attention at Spaceport America. At some point, they will have head-to-head competition at the facility, when Armadillo Aerospace begins offering a very different spaceflight experience on its vertically launched rockets. In either type of launch, the image of commercially available, recreational space flights captures the imagination.

But Spaceport America is about much more than tourism. Other companies were launching unmanned commercial flights from the facility for several years before Virgin Galactic was ready to literally take off. And even Branson's venture will not be limited to carrying passengers. Both the National Oceanographic and Atmospheric Administration (NOAA) and NASA are working with Virgin Galactic to use WhiteKnightTwo to collect data on atmospheric conditions, carry experiments, and possibly launch lightweight satellites.

"Building this spaceport lays the foundation for a whole new industry," Rick Homans said in a 2005 interview. "New Mexico will be the launch pad for America's second space age, centered on private sector innovation and personal spaceflight, and that means new jobs and new opportunities for New Mexicans."

Resources

Belfiore, Michael. *Rocketeers: How a Visionary Band of Business Leaders, Engineers, and Pilots Is Boldly Privatizing Space.* New York: HarperCollins, 2007.

Boye, Alan. *Tales from the Journey of the Dead: Ten Thousand Years on an American Desert.* Lincoln: University of Nebraska Press, 2006.

Homans interview on *The Space Show*, May 11, 2007. Accessed at http://thespace-show.com/detail.asp?q=712.

Kemp, Kenny. *Destination Space: How Space Tourism Is Making Science Fiction a Reality.* London: Virgin Books, 2007.

Epilogue: Critical Mass

It was in New Mexico that Robert Goddard, in 1930, launched the United States' first rocket to soar more than a mile above the ground. In subsequent decades, the US military used New Mexico sites for research in rocketry and human space travel. It took 80 years for those efforts to coalesce into a critical mass. The state is now recognized as a major player in space technology. As a result, it is attracting new opportunities.

A significant example is that in 2010, the Federal Aviation Administration (FAA) selected New Mexico State University in Las Cruces to be the lead agency and headquarters of a new Air Transportation Center of Excellence for Commercial Space Transportation. Other members named to the Center were the New Mexico Institute of Mining and Technology, Stanford University, the University of Florida, the Florida Institute of Technology, the Florida Center for Advanced Aero-Propulsion, the University of Colorado at Boulder, and the University of Texas Medical Branch at Galveston. FAA Administrator Randy Babbitt said, "The research and development efforts will include four major research areas: space launch operations and traffic management; launch vehicle systems, payloads, technologies, and operations; commercial human space flight; and space commerce (including space law, space insurance, space policy and space regulation)."

Also in 2010, NASA's Commercial Reusable Suborbital Research Program (CRuSR) awarded a total of $475,000 to Armadillo Aerospace and Masten Space Systems for a series of test flights of the companies' vehicles. The CRuSR program's goal is to foster the development of commercial, reusable vehicles that can provide regular, frequent, and predictable access to "near-space." A NASA news release said, "The CRuSR awards will fund two flights this fall and one this winter of Armadillo's Super-Mod vehicle from Spaceport America in New Mexico. The first two flights will be to an altitude of approximately 9 miles and the third to approximately 25 miles."

"Now, more than 40 years after Apollo, we are on the verge of a new national vision that can truly inspire the next generation," Spaceport America's Executive Director Rick Homans wrote in the *Albuquerque Journal* in August 2010. "This is not the end of NASA, as some believe, but rather the beginning of a new and exciting space age that will be fueled by innovation and entrepreneurs and set ambitious goals for our national space agency."

Homans admits to thinking that the idea of building a commercial spaceport in the state seemed "a little far-fetched" when he first heard about it, but he has become one of its most dedicated and vocal supporters. "I've always felt like if you get to the point where you really believe that this industry is real and has big potential and that New Mexico has a role, then everything we're doing makes all the sense in the world," he said in a 2010 interview. "But if you don't believe that this industry is coming, then we look pretty off base and kind of nuts. . . . Since 2003 there has not been one event that has created additional doubt or skepticism. In fact, it's been exactly the reverse. There's been steady movement towards this new industry and steady, slow but steady affirmation that this was the right move to make."

Photo Credits

Chapter 1
    All: NASA/courtesy of grin.hq.nasa.gov

Chapter 2
    White Sands: NASA/courtesy of www.nasaimages.org
    Von Braun's surrender: NASA/courtesy of www.nasaimages.org
    Bumper rocket launch: NASA/courtesy of grin.hq.nasa.gov

Chapter 3
    Sam the monkey: NASA/courtesy of grin.hq.nasa.gov
    Ham in training: NASA/courtesy of nix.larc.nasa.gov
    Dittmer and Ham: NASA/courtesy of mix.msfc.nasa.gov
    Isolation cubicle drawing: Courtesy of Holloman Air Force Base

Chapter 4
    Manhigh capsule diagram: US Air Force
    Kittinger by gondola: US Air Force
    Kittinger dropping from gondola: US Air Force

Chapter 5
    Stapp on rocket sled: Courtesy of White Sands Missile Range Museum
    Cobb: NASA/courtesy of grin.hq.nasa.gov
    Scott & Irwin near Taos: NASA/scan by Ed Hengeveld
    Rio Grande Gorge: Wikimedia Commons (photo in the public domain)

Chapter 6
    Chaco Petroglyphs: Photo by James Dale, courtesy of Wikimedia Commons
    Stargazer Balloon: Courtesy of White Sands Missile Range Museum
    VLA: Photo by Loretta Hall

Chapter 7
    Roswell UFO Museum: Photo by Jerry Hall
    Aeroshell: Photo by Loretta Hall
    Scientology Circles: Courtesy of U.S. Geological Survey

Chapter 8
    2006 Rocket Launch: Courtesy of Spaceport America
    2006 Pixel: Courtesy of Spaceport America
    2007 Mod: Courtesy of Armadillo Aerospace
    2008 Mod: Courtesy of Armadillo Aerospace

Chapter 9
    Spaceport America terminal profile: Spaceport America Conceptual Images
    URS/Foster + Partners
    Spaceport America terminal interior: Spaceport America Conceptual Images
    URS/Foster + Partners
    WhiteKnightTwo and SpaceShipTwo flying: Courtesy of Virgin Galactic
    SpaceShipTwo with raised tail wings: Courtesy of Virgin Galactic

Index

A

Aerobee rocket 50, 173
Alamogordo, NM 41, 42, 58, 72, 78, 93, 113,
        123, 141, 154
Aldrin, Buzz 162
Alien encounters 121-22, 124-25, 129-30,
        132-33
Allen, Paul 139-40, 160, 162
Animals
        Cats 47, 54
        Chimpanzees 52--61, 86
        Dogs 47, 57, 68
        Fruit flies 7, 35, 47
        Mice 37, 47, 50
        Monkeys 47-51, 56
        Mutilations 130, 132
Ansari, Anousheh 137, 139
Apache Point Observatory 113-14
Apollo program 41, 92, 101-02, 153, 169
Armadillo Aerospace 141-45, 147-48, 160,
        168-69
Armstrong, Neil 162
Astronauts
        Selection 74, 93-94, 96
        Training 98, 101-03, 152, 161, 163-64

B

Balloon(s) 45-48, 63-83, 109, 112-13, 122-
        24, 126, 129, 162
Beaupre, Francis 75-76, 78, 81
Beeding, Eli 74, 91-92
Binnie, Brian 139
Blossom program 49-50
Boeing 135, 139

Branson, Richard 160-64, 166
Broomstick scientists 37-39
Bumper program 21, 39-40

C

Camino Real 5-6, 156
Cape Canaveral, FL 39, 41, 56-60
Carmack, John 141-45, 147-48, 160
Celestis 158-59
Chaco Canyon, NM 107-09
Clark University 9-10, 12
Cobb, Jerrie 98-100
Corporal rocket 39, 43
cosmic radiation 34-35, 46-48, 50, 63-65, 72

D

Daisy Track 91-92
Diamandis, Peter 137-42, 145
Dittmer, Edward 53, 57, 60, 92
Doolittle, James 24
Dulce, NM 130, 132-33

E

Edwards Air Force Base, CA 41, 85, 102
El Paso, TX 32-33, 42
Enos (chimpanzee) 55-57, 59-61, 63, 90
environmental impacts 166
Excelsior program 74-76, 78-79, 81, 123-24

F

FAA (Federal Aviation Administration) 153-
        54, 169
Flickinger, Donald 98

G

Gates, Bill 140
Gemini program 61, 92, 101
General Electric 33, 39
Gildenberg, Duke 68, 78, 80, 129-30
Goddard, Esther 13-15, 17, 24
Goddard, Robert 9-25, 27-28, 46, 169
gravity 46, 86, 89, 92-94, 101-02
Guggenheim, Daniel 12-13
Guggenheim, Harry 12

H

Ham/Chang (chimpanzee) 52-53, 55, 57-59,
    63, 90
Heavenly Journeys 159
Henry, James 49, 57
Hermes program 33, 42-43
Holloman Air Force Base, NM 47, 49-50, 52,
    54-56, 58-59, 61, 63-64, 68, 70-
    72, 75-76, 81, 85, 88, 91-92, 101,
    112-24, 145
Holloman High Speed Test Track (HHSTT)
    89-92
Homans, Rick 137, 139-40, 148, 152-55,
    157-58, 160, 167-68, 170
Hurd, Peter 14

J

Jornada del Muerto 6
Juarez, Mexico 42-43

K

Kisk, Albert 13, 19
Kittinger, Joe 64-71, 74-82, 88-89, 92, 112-
    13, 124

L

La Paz, Lincoln 127
Las Cruces, NM 30, 32, 37, 111, 126, 136,
    141-42, 147-48, 154, 167, 169
Lindbergh, Charles 12, 14, 137-38
Little Joe rocket 41, 51
Lockheed Martin 135-36, 139, 157-58
Los Alamos National Laboratory (LANL) 97,
    114-15
Los Alamos, NM 115, 118, 126, 132
Lovelace medical clinic 71, 95-97
Lovelace, William Randolph "Randy" 95,
    97-98, 100

Lunar Excursion Module (LEM) 41
Lunar Lander Challenge 143, 145, 147-48,
    151, 160

M

Magdalena Ridge Observatory 116
Manhigh program 64-65, 67-68, 70-74, 76,
    81, 92, 94, 112, 123
Mansur, Charles 13, 15, 19, 25
Mars Rover 118
Masten Space Systems 142, 148, 169
McClure, Clifton 71-74
Melvill, Mike 139
Mercury program 41, 54, 57-61, 81, 96, 98,
    100-01, 113, 117, 158
Microgravity Enterprises 160
Missile Dogs 41
Mogul program 122-23, 126-27
Mojave, CA 139, 167-68

N

NASA (National Aeronautical and Space
    Administration) 51-52, 56, 58-61,
    71, 81-82, 94-96, 98-102, 112-13,
    135-36, 139, 141-42, 148, 151-52,
    157-58, 160, 162, 168-70
National Solar Observatory 113
New Mexico State University 37, 111, 153,
    169

O

Oñate, Juan de 5-6
Orteig prize 137

P

Parachute 15, 18, 20, 23-24, 45, 47, 49-50,
    65, 72, 74-75, 95, 129
    Beaupre Multi-Stage 75, 77, 78, 80-81
Payload Specialties 159-60
Physical Science Laboratory (PSL) 37-38, 111

R

ranchers 29-30, 124, 130, 156
Redstone rocket 38, 41, 57
Richardson, Bill 136-37, 140, 156, 160
Roswell Incident 121-22, 124-25
Roswell, NM 12-15, 21, 23-25, 27-28, 30,
    121-23, 125, 129
Rutan, Burt 139, 160, 162, 166-67

S

Sachs, Henry 13, 20
Scaled Composites 139, 167
science fiction 34, 163-64
Simons, David 47-49, 53, 63- 71, 73, 88, 92
Socorro, NM 6, 116, 118, 127, 129
spaceports
    Other 153-54, 167
    Spaceport America 6, 136-37, 140, 152-54, 156-61, 163-70
SpaceShipOne 139, 153, 160, 162
SpaceShipTwo 162-67
Space shuttles
    Challenger 82, 167
    Columbia 41, 167
space tourism 152-53, 160, 163, 166, 168
Stapp, John Paul 63-70, 74-75, 85-89, 91-92
Stargazer program 112-13
Star Trek 158

T

Tombaugh, Clyde 36-37, 111, 126-27
Trementina, NM 131, 133
Truth or Consequences, NM 77-78, 126, 154, 156
Tularosa Basin, NM 28-30, 37, 154
Tularosa, NM 78, 81

U

UFOs 121-22, 125-27, 129-30, 132-33
UP Aerospace 156-58

V

V-2 rockets 28, 31-36, 38-39, 42-43, 48-50, 85, 109-11
Very Large Array (VLA) 116-18
Viking rocket 42-43, 130
Virgin companies
    Virgin Atlantic Airways 162, 164
    Virgin Galactic 6, 160-64, 166-68
    Virgin Records 161
von Braun, Wernher 30-34, 38-39, 41, 46

W

WAC Corporal rocket 39-40
Ward, Aunt Effie 9-10, 13, 16, 20
weightlessness 46, 49-50, 54, 58-60, 64, 92-93, 101, 152, 164
Wells, H.G. 10, 11
White Sands 28-29
    WSMR 30, 41- 44, 72, 135-36, 154-55
    WSPG 29-33, 35-39, 42, 47, 49-50
WhiteKnightOne 139, 160
WhiteKnightTwo 162, 164-66, 168
Wright Brothers 9
Wright-Patterson Air Force Base, OH 49, 74-75, 96

X

X Prize 137-42, 153, 160, 162
X Prize Cup 140-43, 145, 147-48, 160

Z

Zamora, Lonnie 127-29

ABOUT THE AUTHOR

Loretta Hall has been interested in space travel since her teenage years. She followed closely the early NASA programs: selection of the first astronauts (the Mercury Seven), the suborbital and orbital missions of Mercury and Gemini, the Apollo steps toward a moon landing. She stayed up until 1:15 a.m. in her Seattle home to watch a live telecast of man's first steps on the moon.

Loretta and her family have lived in Albuquerque since 1977. She is a freelance article writer and book author. In 2009, when plans for Spaceport America, the country's only purpose-built commercial spaceflight facility, began moving forward, Loretta was fascinated to discover the important role New Mexico has played in the development of space travel. She decided to herald that unheralded history by writing the only book to document the historic events in the state and the personal stories of the people who accomplished them. She also created a website, *NMSpaceHistory.com*, to supplement the book with news items and additional insights. Loretta is a member of the National Space Society and the Historical Society of New Mexico.

Loretta is the author of *Underground Buildings: More Than Meets the Eye* and is a contributor to the new anthology *Voices of New Mexico*, which is part of the commemoration of New Mexico's statehood centennial.

CPSIA information can be obtained at www.ICGtesting.com
Printed in the USA
243735LV00002B/1-164/P

9 781890 689797